Basic Wiring Techni

Created and Designed by the Editorial Staff of Ortho Books

Project Editor
Cheryl Smith

Writers
Steve George
John Lowe

Illustrators
Angela Hildebrand
Ron Hildebrand

Photographer
Kit Morris

Photo Editor
Roberta Spieckerman

Ortho Books

Publisher
Richard E. Pile, Jr.

Editorial Director
Christine Jordan

Production Director
Ernie S. Tasaki

Managing Editors
Robert J. Beckstrom
Michael D. Smith
Sally W. Smith

System Manager
Linda M. Bouchard

Editorial Assistants
Joni Christiansen
Sally J. French

Marketing Specialist
Daniel Stage

Distribution Specialist
Barbara F. Steadham

Sales Manager
Thomas J. Leahy

Technical Consultant
J. A. Crozier, Jr., Ph.D.

Editorial Coordinator
Cass Dempsey

Copyeditor
Elizabeth von Radics

Proofreader
Karen Stough

Indexer
Elinor Lindheimer

Editorial Assistant
John Parr

Layout by
Cynthia Putnam

Composition by
Laurie A. Steele

Production by
Studio 165

Separations by
Color Tech Corp.

Lithographed in the USA by
Banta Book Group

Photographers
With the exception of the following, all photographs in this book were taken by Kit Morris.

Fred Lyon, back cover TR

Special Thanks
Andrew Kopac

National Electrical Code® and NEC® are registered trademarks of the National Fire Protection Association, Inc., Quincy, MA 02269-9101. Romex® is a registered trademark of GK Technologies, Inc.

Material in tables on pages 20, 24, and 40 is taken from NFPA 70-1990, the *National Electrical Code®*, copyright © 1989, National Fire Protection Association. This material does not reflect the complete and official standard of the National Fire Protection Association on the referenced subject, which is represented only by the code in its entirety.

Address all inquiries to:
Ortho Books
Box 5006
San Ramon, CA 94583-0906

Copyright © 1982, 1993
Monsanto Company
All rights reserved under international and Pan-American copyright conventions.

$$\frac{8 \quad 9}{98 \quad 99}$$

ISBN 0-89721-251-7
Library of Congress Catalog Card Number 92-71349

THE SOLARIS GROUP

2527 Camino Ramon
San Ramon, CA 94583-0906

Front Cover
This double-gang switch box, which will house two 3-way switches, contains wires from three nonmetallic cables. The 2-wire cable (with ground) that enters the bottom brings electricity from the source. Each 3-wire cable entering the top goes to either a light fixture or a 3-way switch in another electrical box.

Title Page
Rough wiring follows a few logical rules and requires only basic tools and a clear head.

Page 3
Kitchens have very specific requirements, whether you are involved in new work or remodeling.

Back Cover
Top left: Residential wiring begins at the service entrance, which consists of the electrical meter, entrance mast, and main disconnect.

Top right: Most wiring devices in a home, such as receptacles, switches, and light fixtures, are contained in an electrical box set within the wall or ceiling. The wires for this metal box have been roughed in and will be connected to a switch after the wall is finished.

Bottom left: With cabinets installed, walls painted, and tile work completed, this kitchen is ready for finish wiring. After the switches, light fixtures, receptacles, and appliances are installed, tested, and inspected, this wiring project will be completed.

Bottom right: This 'convenience center' is part of a Smart House wiring system that integrates standard 120-volt wiring, telephone and intercom cables, low-voltage wiring, TV access cables, and computer links into one modular system.

Basic Wiring Techniques

BEFORE YOU START

No one would deny that electricity is essential to the modern way of life. It does everything from lighting our homes to performing labor once accomplished only by muscle and sweat. Yet few people know much, if anything, about this current that provides such comfort and convenience; flip a switch or push a button and it works, and that's all that matters. Beyond that, most people find electricity confusing, if not downright mysterious, and intimidating because of the potential danger. Yes, there is a danger, but with reasonable caution and understanding, electrical work can be one of the easiest and most satisfying home-repair and remodeling projects. All that the job requires is a basic knowledge of home wiring.

The service entrance for this home has been wired, and the meter is ready to be installed.

UNDERSTANDING ELECTRICITY

Contrary to popular belief, electricity is not difficult to understand. It functions according to mathematical principles and is completely logical. It behaves in a totally predictable way. Understanding that predictability is simply a matter of taking the time to learn the principles behind it, to pick up the jargon, and to master basic wiring techniques.

Wiring by the Book

Home wiring requires few specialized tools, and most of the procedures and materials used are standardized. Armed with this basic knowledge, which will be provided by this book, you will be able to do almost any home-wiring project successfully and safely.

With electricity, a small amount of knowledge can indeed be a dangerous thing, so this book covers all the essentials of residential wiring. It begins with the fundamentals of electricity, including a complete review of the typical home electrical system. This is the time to grasp the basic technical aspects of home wiring and the theory behind those basics. This first chapter also reviews the necessary safety procedures for working with home wiring and provides a glossary of technical terms you will need to understand.

The second chapter presents the step-by-step procedures involved in preparing a wiring project, whether you are wiring an entire house or just adding an electrical outlet. It covers drawing a plan, evaluating whether to use an existing circuit or add a new one, obtaining a permit, and making a materials list. It also discusses tools, safety procedures, and when it might be wise to hire a licensed electrician.

The next three chapters present practical, step-by-step instructions for basic wiring procedures, from rough wiring techniques to installing a new service entrance to the details of finish wiring. Finally, the last chapter examines the important issues of energy conservation, energy generation, and the new technologies related to home wiring.

Everything in this book is explained one step at a time using a combination of text and precise technical drawings. Each step builds upon the previous step. Even if you aren't tackling a project that requires all the steps listed, understanding them all imparts logic to what you are doing. Using this book as a guide, you should be able to do most simple, common electrical-wiring tasks, from setting a box in a new wall to adding a new receptacle to an existing circuit. If you take the time to acquire a clear understanding of electricity, you may even feel able to do advanced projects such as running a new circuit in your home or yard.

From Power Company to Private Home (and Back)

Perhaps electricity seems so mysterious because you can't see it. You can see what it does, and under the wrong circumstances you can feel it, but you can't see it. This invisibility makes it seem much more complicated than it actually is. Simply stated, electricity is the flow of minute charged particles of energy (electrons) through a conductor, generally a wire. The path this energy follows from its creation or generation until it enters a house is both complicated and incredibly efficient.

The path begins at the electric utility company's power plant where electric energy is produced in huge generators powered by water, coal, oil, or nuclear fuel. Typically, this energy is generated at high voltages. Then the power plant uses transformers to further increase its voltage for transmission on high-voltage wires to substations along the line. The actual voltage of each transmission line depends on a combination of factors, including the transmission distance.

At the substation the voltage is reduced for distribution to users. The transformers you see hanging on utility poles or mounted on concrete pads along streets and alleys make the final voltage reduction to 120 and 240 volts for use in the home and are the final link in the intricate distribution system. Along the way, a network of safety devices is used to protect and control the power in the system.

The Basics of Grounding

Grounding is electrical contact with the earth; a ground is a conducted path between electrical equipment and the earth. If you are standing on wet earth or a wet concrete floor that adjoins the earth and you contact voltage, you can become grounded: An electrical current will pass through you, giving you an electrical shock, possibly severe enough to kill you. However, if your home's electrical system, from the service panel through the branch circuits, is properly grounded, it greatly lessens the chance of shock.

One advantage of having a properly grounded neutral conductor in the house is that the voltage to ground will not exceed approximately 120 volts. Typically, voltage to ground in the electrical distribution system ranges from 2,400 to 8,000 volts. The transformer that reduces these high voltages to the 120 volts used in a house has a high-voltage side and a secondary, 120-volt side. Electric power for a home comes from the secondary side. For absolute safety, the distribution system connects the grounded conductor on the high-voltage side of the transformer to the grounded conductor on the 120-volt side. So long as this connection is sound and grounded, an accidental ground on the high voltage feeding the transformer will create a fault that will trip a circuit breaker or open a fuse or otherwise turn off the high voltage and, of course, the 120 volts feeding the house. However, if the common connection

From Power Plant to Home

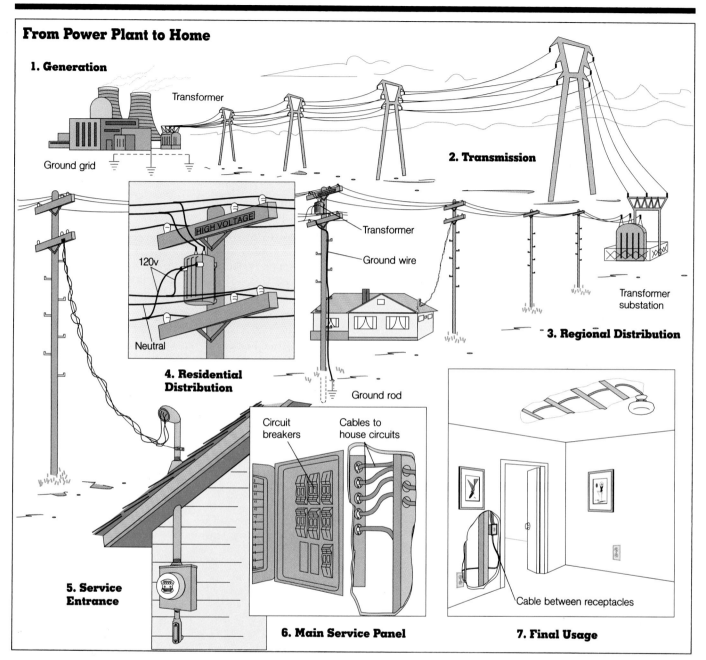

1. Generation

Transformer

Ground grid

2. Transmission

HIGH VOLTAGE

120v

Neutral

4. Residential Distribution

Transformer

Ground wire

3. Regional Distribution

Transformer substation

Ground rod

5. Service Entrance

Circuit breakers

Cables to house circuits

6. Main Service Panel

Cable between receptacles

7. Final Usage

were still in place but the grounding conductor were eliminated, there could be trouble if an accidental ground occurred in the high voltage supplying the transformer. Although the actual result would depend on the primary voltage involved, theoretically the voltage to ground in your house could exceed 8,000 volts.

To further assure proper grounding, all house receptacles, in addition to being connected to a hot wire (most often black or red) and to a neutral wire (white or grayish white), must be connected to a ground wire (always bare copper or green) that joins to a grounded connection in the service panel.

This connection is made either through a grounding conductor in the cable feeding the receptacle or through the sheathing of metal-jacketed cable or metallic conduit used in the system.

A washing machine serves to demonstrate how this grounding inside the house protects you. When the machine is plugged in and turned

on, only the hot and neutral wires to which the washer is connected normally carry current. The machine and its drive motor are constructed in such a way that their wiring is insulated or physically isolated from the machine's metal frame. For further protection, the washer's frame also is connected to ground through the green wire in its cord and the U-shaped prong in its plug.

However, if excess heat or some other cause breaks down the drive motor's insulation, its frame and the washer's frame will become energized when the machine is turned on. If the electrical system in the home is properly grounded, the high current flowing through the motor to ground will open the circuit breaker or fuse protecting the circuit supplying power to the washer and no damage will occur. However, if the grounding prong in the plug is missing or the receptacle isn't properly grounded, and you happen to touch the machine and the properly grounded clothes dryer next to it at the same time, the fault current will flow from the washer through you to ground and a fatality could occur.

Electricity in the Home

From the transformer, the power company brings electricity into a home via overhead wires or underground cables. Underground service is aesthetically less intrusive, but there is no difference in the quality or safety of the electrical service you receive.

Homes built after 1940, and pre-1940 homes that have had their electrical service modernized, will have three-wire service. This means there are three wires, called service-entrance conductors, connected to the service-entrance equipment of the home. Two of these wires are "hot" (carrying current), the third one is the neutral conductor.

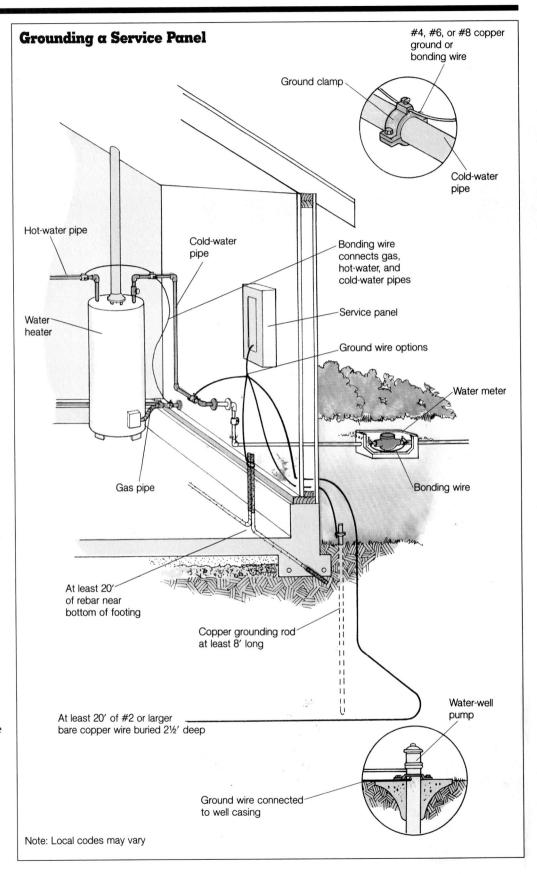

Grounding a Service Panel

#4, #6, or #8 copper ground or bonding wire

Ground clamp

Cold-water pipe

Hot-water pipe

Cold-water pipe

Bonding wire connects gas, hot-water, and cold-water pipes

Service panel

Ground wire options

Water heater

Water meter

Bonding wire

Gas pipe

At least 20' of rebar near bottom of footing

Copper grounding rod at least 8' long

At least 20' of #2 or larger bare copper wire buried 2½' deep

Water-well pump

Ground wire connected to well casing

Note: Local codes may vary

The three-wire system provides a home with both 120-volt and 240-volt. The neutral wire and either hot wire combine to supply 120 volts for such items as lamps, radios, food processors, and wall receptacles. When 240-volt service is necessary, to operate a major appliance such as a range or clothes dryer, for example, both hot wires combine with the neutral wire to supply the required electricity. However, note that water heaters and large air conditioners also require 240 volts, but they are connected only to hot wires. They don't have clocks, timers, lights, or other appurtenances that require 120 volts for their operation.

Older, unmodernized homes usually have two-wire service and as few as two fuses. This means they have one hot wire and one neutral wire supplying only 120 volts. This service is wholly inadequate by modern standards, especially if the home is equipped with many large appliances.

The Service Entrance

The electrical distribution system within the home—an intricate but logical and efficient system—is centered in the home's service-entrance panel. Three wires bring power from the 120-volt side of the distribution transformer through the electric meter and then into the service-entrance panel. If the house has underground service, these wires, called the service lateral, rise through the entrance conduit that comes out of the ground alongside the foundation. They then enter the meter, which is mounted on the outside of the

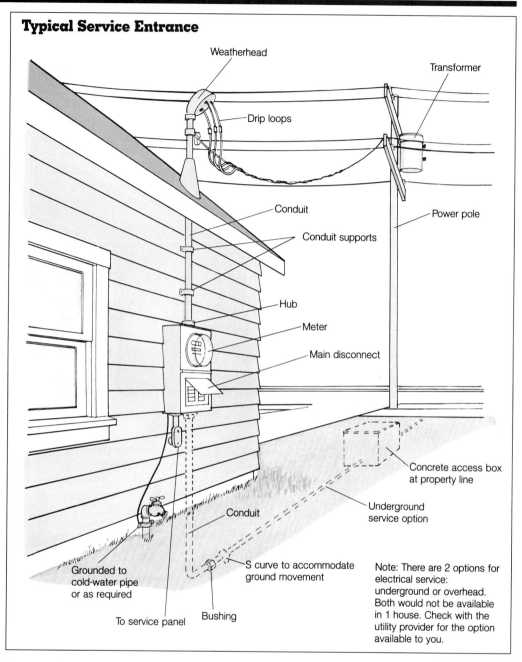

Typical Service Entrance

Weatherhead

Transformer

Drip loops

Conduit

Conduit supports

Power pole

Hub

Meter

Main disconnect

Concrete access box at property line

Underground service option

Grounded to cold-water pipe or as required

Conduit

S curve to accommodate ground movement

To service panel

Bushing

Note: There are 2 options for electrical service: underground or overhead. Both would not be available in 1 house. Check with the utility provider for the option available to you.

house, and continue on to the service-entrance panel inside. If the house has overhead service, the three wires, called the service drop, extend from the secondary side of the distribution transformer to a point near the service-entrance head at the top of the service conduit. Here they are spliced to the service-entrance conductors, which enter this head, drop down through the conduit into the meter, and continue on into the service-entrance panel.

The Electric Meter

The electric meter is attached to your house but it belongs to the power company. All the electricity entering the house first passes through this meter so the power company can measure the exact amount of electricity you use. This measurement is expressed in kilowatt-hours (*the number of kilowatts × the hours of usage*). To learn how to read your meter, see page 12.

The Service Panel

It would be convenient if all service equipment and every service arrangement were exactly the same, but they're not. There are many types of panels and many acceptable ways to locate them in relationship to the meter. The common denominator among panels is that each must be grounded and contain a main disconnecting means and a main overcurrent-protection device.

The main disconnect is wired to the load side of the meter and can mechanically turn off all the power in your house. The main overcurrent-protection device can automatically turn off all the power to your house. Most contemporary panels combine the main disconnect and the main overcurrent-protection device in the main circuit breaker, usually located at the top of the panel just above the branch-circuit breakers. However, you should be aware that some local electrical codes require the main disconnect to be located outside, away from the service panel. You will have to use the type required by your community's code. If your home has a fuse box, the main disconnect will consist of a pair of fuses mounted in a pull-out, which is removed to cut the power. With some old split-bus–type panels you may need up to six separate hand movements to disconnect all the incoming power. Be sure you know what type of panel you have and exactly what steps you must take to disconnect the power completely.

Service Panels

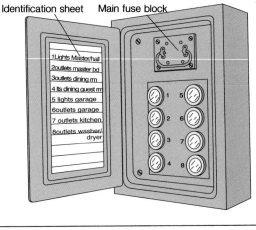

Identification sheet   Main fuse block

Expansion blanks

The service-entrance panel also contains the grounding-electrode conductor, an important wire that grounds the entire house. The neutral wire inside the service panel goes directly to the neutral bus bar. All the neutral wires of the branch circuits connect to terminals on this solid metal bar that is bonded (electrically connected) to the cabinet. The grounding-electrode conductor connects the neutral bus bar to a metal ground rod driven into the ground beyond the foundation of the house. If the water system for the home includes more than 10 feet of underground metal water pipe (including metal well casing bonded to the pipe), the water system also is connected to the neutral bus bar, as specified by the *National Electrical Code* (*NEC*). Working together, the ground rod and the water pipe form the grounding-electrode system that provides a safety path to earth for the electrical system of the home.

The Role of the *NEC*

The electrical practices and procedures presented in this book are based on the *National Electrical Code*, hereafter referred to as the *NEC* or the Code. Sponsored by the National Fire Protection Association, the *NEC* establishes national minimum safety standards for the installation of electrical wiring and equipment. The Code has one purpose: to safeguard people and property from hazards arising from the use of electricity. It is designed for mandatory application by any authority having legal jurisdiction over electrical installations. By conforming to the *NEC*, a book such as this is able to provide accurate instructions to a nationwide audience of readers, even though they are governed by local electrical codes that can vary from community to community.

Most of these local electrical codes, whether city, county, or uniform state codes, are based on the *NEC*, but they may vary in significant ways; they may be more stringent or they may not be as up-to-date. The *NEC* revises the Code every three years. Since all electrical work in the home must be done according to code, you must check out your local electrical code before you even begin planning your project. To do this, contact the building-inspection department for your community, usually located in city hall, and ask for a list of electrical-code requirements. Also ask if the code follows the *NEC*. If there are significant deviations, what are they? Some local codes may regard the *NEC* as an acceptable alternative. Whatever the situation, all your electrical work must be done according to the code used by the authority that has jurisdiction over your residence.

Even though you have a copy of your local electrical code, you should still familiarize yourself with the *NEC*. The best way to do this is to obtain a current guide to the *NEC*, because the Code itself is quite complex.

Circuit Plan of a Typical Home

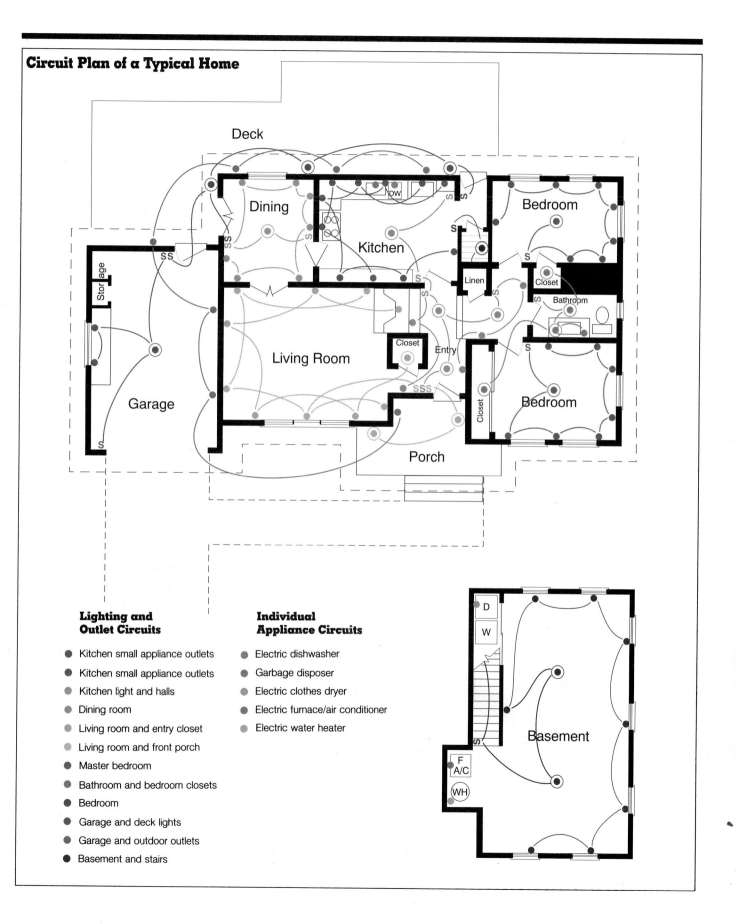

Lighting and Outlet Circuits

- Kitchen small appliance outlets
- Kitchen small appliance outlets
- Kitchen light and halls
- Dining room
- Living room and entry closet
- Living room and front porch
- Master bedroom
- Bathroom and bedroom closets
- Bedroom
- Garage and deck lights
- Garage and outdoor outlets
- Basement and stairs

Individual Appliance Circuits

- Electric dishwasher
- Garbage disposer
- Electric clothes dryer
- Electric furnace/air conditioner
- Electric water heater

How to Read Your Electric Meter

You are charged for the electricity you use in kilowatt-hours (kwh). *Kilo* means thousand, so a kilowatt is a thousand watts. These kilowatts are measured by the electric meter attached to the house, which runs continuously and registers your use on the four or five dials on its face. Some meters have a counter similar to the mileage indicator on a speedometer, but most have dials that resemble clocks, with numbers running from zero to nine. Look closely at these dials and you'll see their numbers alternate between running clockwise and counterclockwise.

To read your meter, start with the left-hand dial and work to the right, jotting down the numbers as you go. If the hand lies between two numbers, always use the lower number even if the hand is almost on the higher number. However, if the hand is directly on a number, check the dial immediately to the right. If the hand on that dial has reached or just passed zero, use the number indicated on the previous dial; however, if the hand of the dial on the right has not reached zero, use the next lowest number on the previous dial.

Some homeowners like to take their own meter readings every once in a while in order to check the accuracy of the power company's readings. To do this, write down the meter reading at the same time the company reads the meter. Do this again the following month, and subtract the first reading from this second reading. Then compare your consumption with that stated on your electric bill for that month. If there is a significant discrepancy, contact the power company.

The Branch Circuits

Branch circuits within the home are formed by the wiring that connects electrical outlets (receptacles) and permanently connected appliances to the service-entrance panel. Each individual circuit has its own hot wire and neutral wire as well as an overcurrent-protection device capable of turning off power to the individual branch circuit.

In the panel, two hot wires from the main circuit breaker energize two bus bars constructed of heavy metal, usually copper or aluminum. Each branch circuit is connected to one or both of these hot bus bars via individual overcurrent-protection devices, either circuit breakers or fuses. A 120-volt circuit has one hot wire coming from a single-pole circuit breaker, that is, a circuit

breaker connected to one hot bus bar, and a neutral wire or conductor that is connected to the neutral bus. A 240-volt circuit has two hot wires connected to a double-pole (or two-pole) circuit breaker, that is, a circuit breaker that is connected to both hot bus bars. However, this 240-volt branch circuit may or may not have a neutral wire connected to the neutral bus; that's because 240-volt circuits for water heaters and large air conditioners do not require neutral wires (see page 9 for explanation).

It is important to remember that all neutral conductors terminate at the neutral bus.

Most homes are equipped with a variety of branch circuits. The *NEC* recognizes three types.

General-purpose Branch Circuits

These 120-volt circuits supply a number of receptacles used for lighting and small appliances. Some designers and installers prefer to assign a circuit only to lighting or only to receptacles. While this practice is not an *NEC* requirement it may be a local code requirement in some communities. It may also be common practice even where it is not an actual code requirement. This approach is advisable for houses that include many permanently installed light fixtures.

Small-Appliance Branch Circuits

These 120-volt circuits supply power to receptacles to which small appliances such as toasters and food processors are to

be connected. Typically, these circuits are restricted to kitchens and eating areas. A minimum of two such 20-amp circuits are routinely installed in kitchens built within the last 30 years.

Individual Branch Circuits

These circuits supply only one appliance. There are two types: 120-volt circuits and 240-volt circuits. The 120-volt circuits service appliances such as dishwashers, food disposers, trash compactors, and washing machines. The 240-volt circuits are for appliances and equipment with greater power requirements, such as clothes dryers, ranges, water heaters, and air conditioners.

UNDERSTANDING ELECTRICAL CAPACITY

The wires used for conducting electricity offer low resistance to the flow of electric current, but the friction produced by the electrical flow produces heat. A small amount of heat is normal, but when it reaches a level where the insulation around the wire begins to fail the current must be reduced or turned off to avoid a fire.

Vocabulary of Electricity

To understand electrical capacity, you need to know the terminology that describes the various components of electric power. These words are *amperes*, *volts*, and *watts*. They describe common functions and are interrelated in precise ways.

To comprehend this relationship, think of electricity as similar to water flowing through a hose. Like the water, the flow of electricity must be contained, in this case, in a wire. The moving electricity is a current, called amperes, and it moves under pressure, called volts. When you multiply these two elements, *amperes (current)* × *volts (pressure)*, you get the number of watts, which is a measurement of the energy being used, or to state it another way, a measurement of the work being done.

Amperes, typically called amps, is the measurement of the number of electrons flowing past a given point each second. It takes 6.28 billion billion electrons passing a given point each second to produce 1 amp, which is barely enough to ring your doorbell.

Volts measure how much pressure is being used to push electricity through the wire. Thus, the amount of current flowing through a given circuit depends on the amount of voltage being applied to that circuit. Because voltage is a measurement of pressure, it is subject to resistance. This resistance can be measured as well; it is stated in ohms. Voltage alone cannot produce work, but when it is applied to a circuit, a current flows and work is produced.

The voltage arriving at a house can fluctuate from 114 to 126 volts. That is why you may see references elsewhere to household voltages other than 120 and 240. Like the *NEC*, this book refers to voltage as being 120 and 240, the medians of the fluctuations.

Watts measure the amount of power being used at any given moment. Taken separately, neither amperes nor volts can tell you how much power is being used; they work together to produce power, which is expressed in watts. To calculate watts, simply multiply volts by amps. For example,

a standard light bulb drawing ½ ampere from a 120-volt circuit uses 60 watts of power (*120 volts* × *0.50 amps = 60 watts*). Because watts are the result of an equation, variable factors can produce the same wattage. For example, that 60 watts in the light bulb is equivalent to the 60 watts produced when a car headlight draws 5 amps from a 12-volt battery (*12 volts* × *5 amps = 60 watts*).

Knowing these concepts and the simple *amps* × *volts = watts* formula can help you in many ways as you undertake a wiring project. For instance, if you need to install a new electric clothes dryer, you will have to know what size wire to use and what size circuit breaker to choose to protect the wire. If the clothes dryer is 240 volts and is rated at 7,200 watts, you could calculate the required amperage by changing the $A \times V = W$ formula to $W / V = A$, or $7200 / 240 = A$. The answer is 30, so you would use a wire size able to carry 30 amps, which would be No. 10 copper wire, protected by a 30-amp breaker.

Preventing Circuit Overload

Knowing how amps, volts, and watts interrelate can help you prevent the dangerous condition known as circuit overload. When a circuit breaker trips or a fuse blows, it means too much current is being drawn on that circuit. You can calculate the current a circuit will draw simply by adding up the watts of all the lights and appliances being used on it at the

same time. (The wattage is stated on each of these items.) Divide the total watts by 120 volts to get the amperage being drawn on that circuit. The resulting value should not exceed the amperage rating marked on the corresponding fuse or circuit-breaker handle. For example, 1,800 watts is the maximum power that can be delivered by a 15-amp, 120-volt circuit (*15 amps* × *120 volts = 1,800 watts*). Exceeding 1,800 watts will overload the circuit. Until that overload is relieved, the overcurrent-protection device for that circuit will keep tripping or blowing, protecting you and your property. If you ignore these warnings and replace a blown fuse with one with a higher amperage rating than indicated for the circuit, you create potential conditions for overheating wires.

Of course, this situation cannot occur in a house where the circuit breakers or fuses properly correspond to the wire sizes being used. Whichever overcurrent-protection device you have would open the circuit long before there was any danger, warning you something was wrong. Don't ignore that warning.

Just as a wire must be protected by a properly sized circuit breaker or fuse, the wire itself must be large enough to handle the load being served by that circuit. This point is crucial and is discussed in greater detail in the "Rough Wiring" chapter.

SAFETY

That old saying, "safety first," may be a cliché, but it's also true. Although electricity does marvelous work for us, under the wrong conditions it can be dangerous. You should not work with electricity unless you are prepared to treat it with the caution and respect it deserves. Use common sense and obey a few simple safety rules when you work.

Rules for Working With Electricity

• The most important rule is: *Always shut off the electric power to the circuit on which you plan to work.* Never attempt to touch wires or energized equipment before you kill the circuit. To make this disconnect, go to the service panel and flip the circuit breaker to *off* or unscrew the fuse that protects that circuit. Lock the box or leave a note to alert others that you are working on the circuit. Then, before you touch any wires, use a voltage tester to make sure it is off. If you don't have a voltage tester, turn on a light or appliance connected to the circuit before you make the disconnect; the light will go off when you shut down the circuit. As an added precaution because you can't be too careful, tape the circuit breaker in the *off* position or take the fuse with you, and post a sign on the box to warn others you are working on the circuit.

• Electricity and water do not mix. *Never stand on a wet or damp floor when working with electricity.* Put down dry boards or a rubber mat on which to stand while you work. Also, never work with electricity if *you* are wet. Change into dry clothes, including shoes, before you begin work.

• *Never touch any plumbing pipes or fixtures, radiators, or metal duct work while working with electricity.* If you touch a hot wire and a water pipe, for example, an electric current could flow through your body.

• *Never touch the wires supplying your service panel.* They are still live even if you have pulled the main disconnect. Keep aluminum ladders away from these lines, too.

• When your electrical work is completed, turn on the power and immediately check your work with a voltage tester. The tester should light when a connection is made between the hot wire and the grounded box; it should not light when a connection is made between the neutral wire and the grounded box.

• Use tools with rubber- or plastic-covered handles.

• Wear safety goggles or glasses and, when feasible, gloves.

• Use only one hand when working on service equipment, such as a circuit-breaker panel. Using two hands creates the risk of completing a circuit and having current flow through your body if you inadvertently touch a hot conductor.

• Perform all work in accordance with the *National Electrical Code* and local requirements, and obtain the proper permit from your local building department when required.

Rules for Using Electricity

To use electric power safely in your home, follow these household safety tips.

• Never use a circuit breaker or fuse with an amperage rating higher than that specified for the circuit.

• Make it a practice to cover all unused receptacles so metal objects cannot accidentally be poked into the slots. Even better, install safety covers on all receptacles.

• To prevent fires, never run a lamp, appliance, or extension cord under a carpet or rug or through a doorway or other traffic lane where wear and tear can result in frayed insulation, exposing bare wires to flammable materials.

• Likewise, don't plug more watts into an extension cord than it is rated to handle. If you do, the resulting overheating will destroy the insulation in the cord, again exposing bare wires to flammable materials. To determine how much electricity an extension cord can safely support, multiply the amperage rating of the cord (13 amps for a No. 16 cord, for instance) by 120 volts; the total is the number of watts the cord can handle safely. Then add up the total wattage of the items you want to plug into the cord; if that total exceeds the total permitted watts, something will have to be unplugged. Always unplug a lamp or appliance before attempting any repairs to it.

• To safely remove a power cord from a receptacle, grip the plug firmly with your thumb and forefinger and pull it out. Never yank a plug from a receptacle by pulling on the cord.

• When using an adapter for a three-pronged plug, be sure you properly ground it by connecting it to the screw that holds the wall plate to the receptacle.

• When you discover a cracked plug or a frayed cord, repair it immediately. Better yet, replace it with a new unit.

• Protect all bathroom and kitchen receptacles near the sink with a ground fault circuit interrupter (GFCI). All outdoor, garage, and basement receptacles should also be protected with a GFCI. The *NEC* requires GFCIs to be installed in these areas in new work. You can buy units that wire into regular receptacles in old work in order to provide this protection for your home.

• Install smoke detectors in your house and garage.

Safety Principles

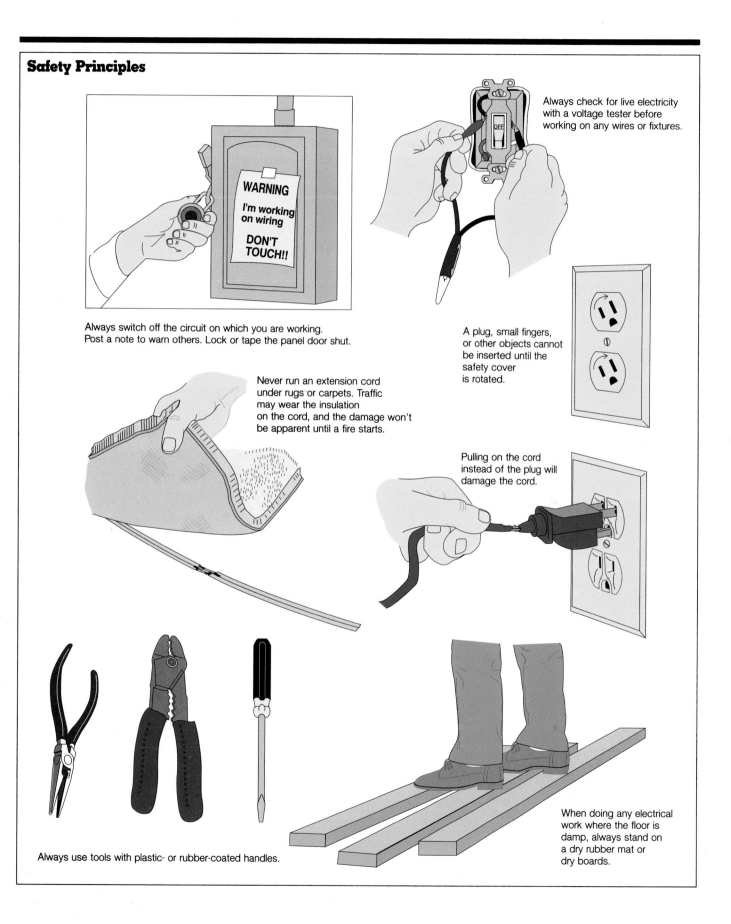

Always switch off the circuit on which you are working. Post a note to warn others. Lock or tape the panel door shut.

Always check for live electricity with a voltage tester before working on any wires or fixtures.

A plug, small fingers, or other objects cannot be inserted until the safety cover is rotated.

Never run an extension cord under rugs or carpets. Traffic may wear the insulation on the cord, and the damage won't be apparent until a fire starts.

Pulling on the cord instead of the plug will damage the cord.

Always use tools with plastic- or rubber-coated handles.

When doing any electrical work where the floor is damp, always stand on a dry rubber mat or dry boards.

PLANNING THE WORK

Little, it seems, begins without a plan. This is certainly true when it comes to electrical work. Before you start any project, you need to map out the work to be done. This chapter guides you step-by-step through the techniques used to plan a wiring project of any size and draw it on paper. You will learn how to determine the number of fixtures the house needs, organize them into branch circuits, calculate the service load, and select and locate the service panel. Although this book approaches the job from the perspective of wiring a whole house, the same principles apply to smaller projects, such as minor repairs or installing a new electrical outlet. Understanding the whole system will help ensure successful completion of even small jobs.

Take plenty of time to consider all the options available. Hours invested in careful planning are well spent and will save both time and money in the long run.

DRAWING A PLAN

If you do the job yourself, you make all the decisions. If you hire an architect or builder to develop your project, use the situation to learn all you can about how these professionals plan an electrical system. Ask all the questions you want and make your wishes known. You determine the details of your wiring plan; the professional interprets them.

The Basic Floor Plan

Begin with a copy of the floor plan of the house or addition. This is the starting point whether you are drawing the existing circuitry or laying out a new electrical system for a whole house or a major addition. Spread the floor plan in front of you and refer to the requirements in the Receptacle Guide section below to chart the receptacles for each room. Use the appropriate symbols shown in the symbol legend (see page 20). Your power company can help you develop a plan. In fact, it must be involved in the process if you are planning an increase in service or installing a new service-entrance panel. Its technical advice is free.

Receptacle Guide

• Receptacles are required in all finished habitable spaces in the home. They must be installed so that no unbroken wall space, when measured horizontally along the floor line, is more than 6 feet from a receptacle. Any wall space 2 feet or wider requires at least one receptacle. Wherever possible, the receptacles should be spaced equal distances apart. A receptacle mounted in the floor close to a wall is considered to be on the wall.

• Receptacles in kitchens, pantries, informal eating areas, and dining rooms must be served by two or more 20-amp appliance branch circuits. No other receptacles can be attached to these circuits.

• A receptacle must be installed at each countertop space wider than 12 inches, and no point along the countertop can be greater than 24 inches from a receptacle. Island and peninsular countertops wider than 12 inches must have one receptacle for each 4 feet of length. Countertop spaces separated by sinks, refrigerators, ranges, wall ovens, or cooktops are considered separate spaces.

• At least one wall receptacle is required next to the sink in each bathroom.

• A receptacle accessible at grade level must be installed outdoors.

• At least one receptacle is required in each laundry area. This must be on a separate circuit; no other receptacles can be located on it.

• An attached garage must have at least one receptacle.

• A basement requires at least one receptacle in addition to that required for the laundry area if it happens to be located in the basement.

• At least one receptacle is required in hallways 10 feet or more in length.

Lighting Requirements

Next, draw in the required light fixtures and their switches, using the appropriate symbols. Note that in all habitable rooms except kitchens and baths, a switched receptacle (usually the top half of a duplex receptacle) satisfies the requirement that each of these rooms have at least one switch-controlled light fixture. Kitchens, bathrooms, stairways, hallways, and attached garages require switch-controlled permanent light fixtures, as do outdoor entrances or exits. The same is required for attics, under-floor spaces, utility rooms, and basements if these areas are to be used for storage or to house equipment that requires servicing.

Individual Branch Circuits

Now, mark the locations for the individual branch-circuit receptacles: oven, range, cooktop, dishwasher, food disposer,

water heater, heating and air-conditioning equipment, and the like. These appliances must all have their own individual circuits that serve only the one appliance and no other appliances, receptacles, or fixtures. Identify each as shown in the symbol legend on page 20. Decide where to locate the service equipment. When a new service is involved, this step must be done in cooperation with the power company. A sensible location for the floor plan shown is on the rear wall of the garage because of its proximity to the larger electrical loads of the house.

Other Options

These are the basic requirements. There are many other items that should also be investigated for incorporation into the electrical design. Smoke detectors are required in most municipalities; even if not required they should be considered a must. *NEC* requirements for lighting switches are minimal, as discussed above, so your plan should also include a convenient light-switching scheme. Think about mood and accent lighting, ceiling fans, exhaust fans, supplemental heating in bathrooms and dressing rooms, and so forth. Remember to include doorbells and, possibly, a garage-door opener receptacle. A communications system, including telephone, intercom, and telecommunication capabilities, are other items to consider. Draw in and identify any options you want included in your electrical plan. None are shown on the sample floor plan.

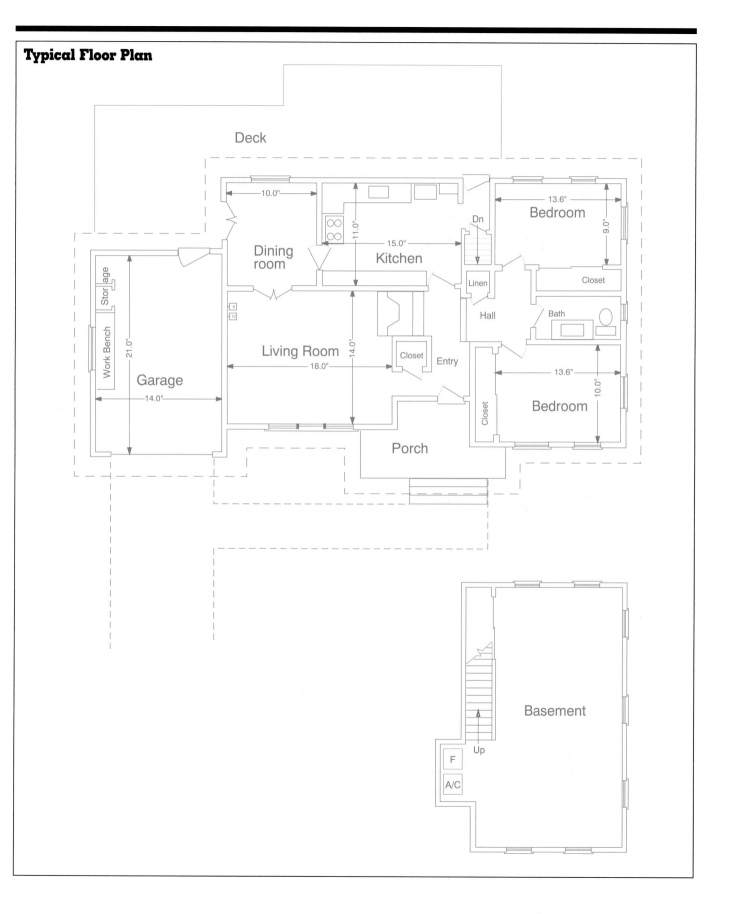

Common Electrical Symbols

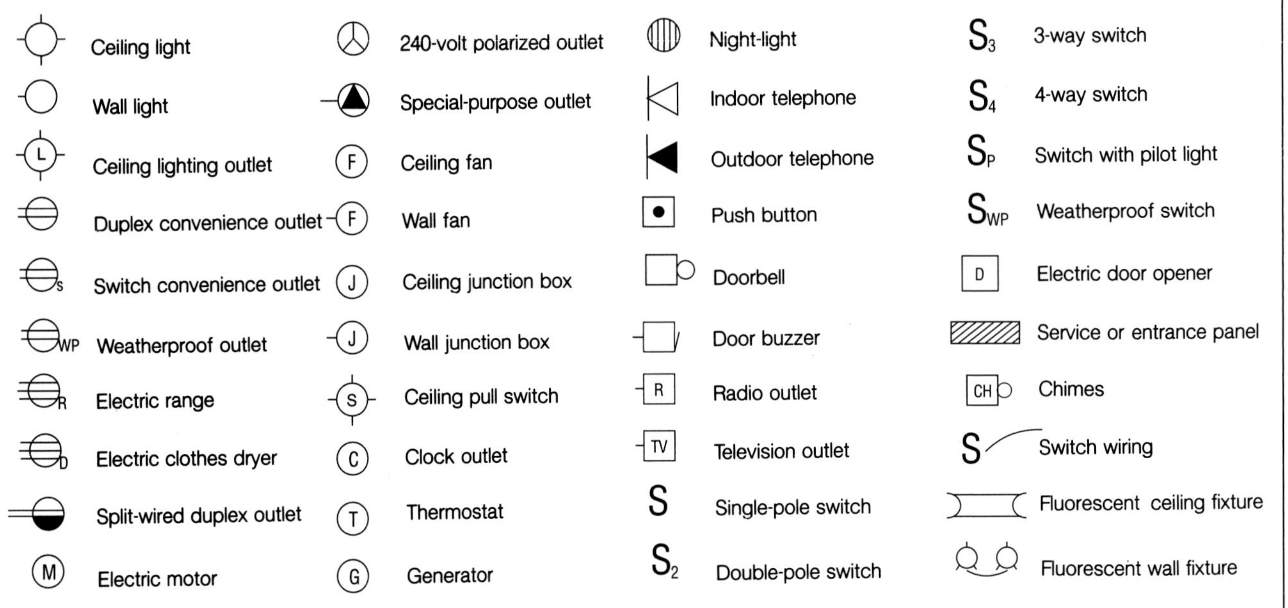

- Ceiling light
- Wall light
- Ceiling lighting outlet
- Duplex convenience outlet
- Switch convenience outlet
- Weatherproof outlet
- Electric range
- Electric clothes dryer
- Split-wired duplex outlet
- Electric motor
- 240-volt polarized outlet
- Special-purpose outlet
- Ceiling fan
- Wall fan
- Ceiling junction box
- Wall junction box
- Ceiling pull switch
- Clock outlet
- Thermostat
- Generator
- Night-light
- Indoor telephone
- Outdoor telephone
- Push button
- Doorbell
- Door buzzer
- Radio outlet
- Television outlet
- Single-pole switch
- Double-pole switch
- 3-way switch
- 4-way switch
- Switch with pilot light
- Weatherproof switch
- Electric door opener
- Service or entrance panel
- Chimes
- Switch wiring
- Fluorescent ceiling fixture
- Fluorescent wall fixture

Sizing a Service Panel

Lighting load 1,220 sq ft @ 3va/sq ft	3,660 (volt-amperes)
Small appliance load (2) 20amp @ 1,500va	3,000
Laundry circuit (1) 20amp @ 1,500va	1,500
Furnace blower	1,000
Wall-mounted oven	6,500
Cooktop	5,100
Dishwasher	1,000
Food disposer	1,000
Clothes dryer	5,000
Total other load	27,760
1st 10,000va @ 100%	10,000
Remainder @ 40%	7,104
Total net other load	17,104
Air-conditioning load	3,600
Total net load (volt-amperes)	20,704
Service rating amperes (total net load/240 volts)	86 amps

Based on material from NFPA 70-1990, the *National Electrical Code®*, copyright © 1989, National Fire Protection Association, Quincy, MA 02269.

Sizing Your Service

Choosing the size of your service panel is a three-step process. The first step involves computing the electrical service load based on *NEC* Article 220, Branch Circuit and Feeder Calculations. The outside dimensions of the home (*length × depth*) are used to determine its square footage. For the sample floor plan, the finished area is 1,220 square feet. Also note that the heating plant and water heater are fueled with natural gas; everything else is electrically powered.

Using one of the *NEC*'s optional calculations for this example, the adjacent worksheet was developed. (Note: Values shown are volt-amperes [va]. To obtain volt-amperes multiply *volts × amps,* using the values shown on the equipment nameplates. If a nameplate lists wattage only, use this instead of volt-amperes.)

This indicates that a 100-amp service will be sufficient for this electrical system. It won't, however, provide much capacity for future expansion, an important consideration in these times of large appliances that greatly increase the load on a home electrical system. A larger panel might be a wiser choice in the long run.

Now the second step: Count the number of circuit-breaker spaces or fuse holders necessary to accommodate the number

Electrical Plan

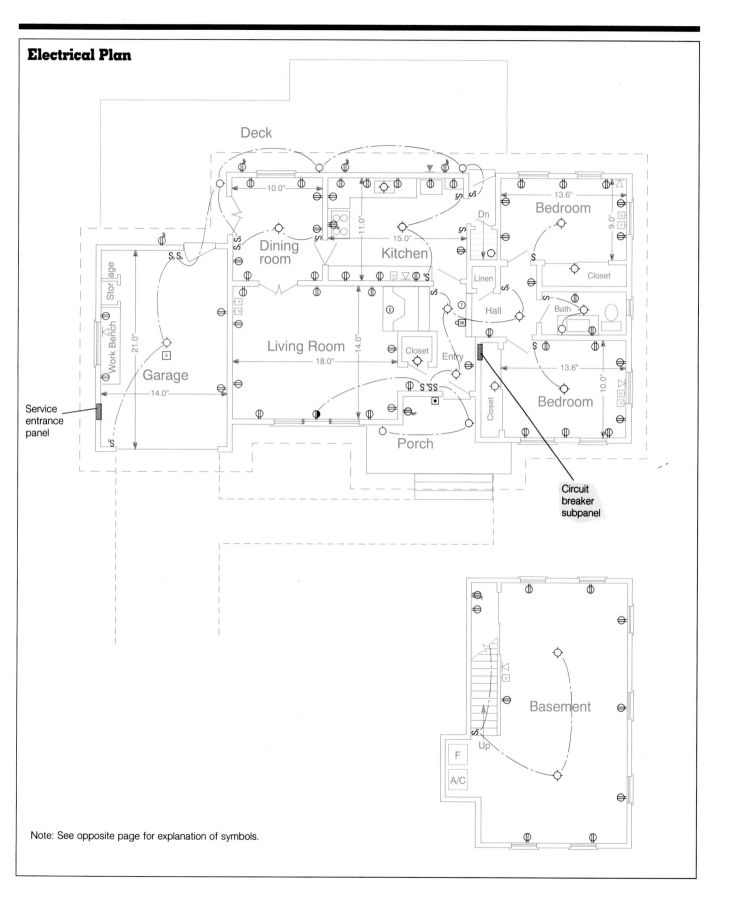

Note: See opposite page for explanation of symbols.

Special Systems Plan

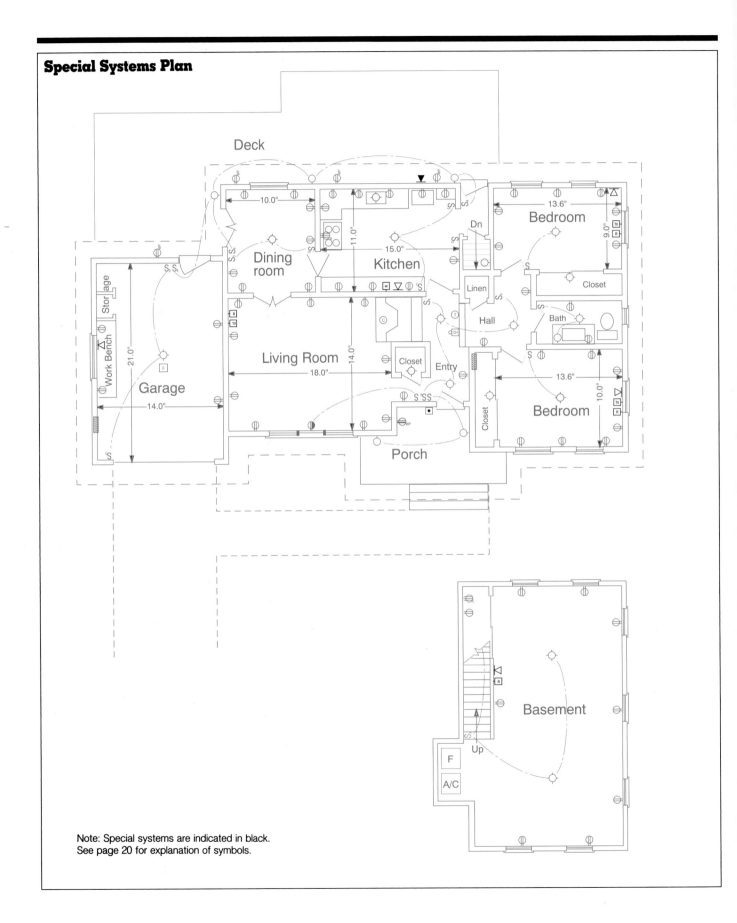

Note: Special systems are indicated in black.
See page 20 for explanation of symbols.

of overcurrent-protection devices needed in the system.

A study of the sample floor plan shows that there are 46 fixtures that are connected to the general-purpose branch circuits. These fixtures are intended to serve the general lighting needs of the home, which were found to be 3,660 volt-amperes. Dividing this by 120 volts shows that 30 amps total, or two 15-amp circuits, could provide sufficient power for the general lighting needs. A much better approach is to divide the number of receptacles by 8, round off to the next higher number, then plan for this many general-purpose branch circuits. In this house, six will be used. Other single-pole (120-volt) circuits requiring their own circuit breakers are those for the food disposer, dishwasher, laundry, and furnace. Two additional breakers are necessary for the small-appliance branch circuits, bringing the total 120-volt branch circuits to 10.

Next note that two-pole (240-volt) circuit breakers are needed for the oven, cooktop, dryer, and air conditioner. A total of four two-pole circuit breakers require eight spaces.

Add them up and a total of 18 spaces are needed in the panel. Generally, 100-amp service panels provide no more than 20 spaces. This is close if

you are considering possible future expansion. It also won't provide for separate lighting and outlet circuits, which are a common practice or even a local code requirement in some communities where large, expensive homes with a lot of permanently installed lighting are built. The wisest course of action would be to select a higher capacity panel which would have more spaces available.

Finally you must choose how the panel will be mounted: flush mounted (recessed in the wall) or surface mounted. Some service panels can be installed either way, but others require surface trim for surface mounting and flush trim for mounting in a wall. It is your choice. Use whichever you will find more convenient. For some years now, the trend has been toward using circuit breakers rather than fuses; new service equipment employing fuses is impossible to find in most areas, so you most likely will have no choice in this matter.

There are some points to keep in mind when locating the service panel. It must be easily accessed, with at least 3 feet of clearance in front of it, and sufficient elbow room on each side. These requirements generally eliminate closets as an option. The panel should be at eye level, and all wiring *must* be enclosed, either in walls or in conduit.

Circuit Breaker Load/Assignments

MAIN
150 AMP

1	Air conditioning	2	Clothes dryer
3	Air conditioning	4	Clothes dryer
5	Oven	6	Cooktop
7	Oven	8	Cooktop
9	Small appliances	10	Dishwasher
11	Small appliances	12	Food disposer
13	Laundry	14	Furnace
15	Lights and Plugs	16	Lights and Plugs
17	Lights and Plugs	18	Lights and Plugs
19	Lights and Plugs	20	Spare
21	Lights and Plugs	22	Spare
23	Spare	24	Spare

Service panels are often placed in the same area as the service entrance. There are even more requirements in this case.

First, work with the power company. It may not be possible to reach your service from its distribution system without resorting to some extraordinary means such as setting another pole, for instance. This is a common occurrence in rural areas, and the customer pays for setting such new poles, as much as $1,200 each pole. You want to mount the panel as near as possible to the point where the service-entrance conductors enter the house, and it must be in a readily accessible space with reasonable clearance around and in front of it. The top of the panel

should be at eye level, approximately 5 feet above the floor.

Don't forget to consider aesthetics as well as more pragmatic issues when you are deciding where to locate the service panel and service entrance. You would not want this vitally necessary but rarely attractive piece of equipment to detract from the visual appeal of your home. Many service entrances are tucked away at the rear of the house, or even in specially created alcoves, so that they are not generally visible.

If you have dogs or other animals that are free to roam the yard, you may want to consider placing the service entrance and electric meter where they can be seen from an adjacent yard. Otherwise, you can arrange with the power company to read your own meter and post the readings once a month.

PLANNING THE WORK

Once you have drawn your floor plan, you are ready to plan which fixtures will be linked and where wire will be run, as well as what materials to use. Take your time and consider your options. Careful planning now will minimize work later.

Wire Sizes

	Cable Assemblies		Individual Conductors	
AWG	14	15 amps		
	12	20		
	10	30		
	8	40		
	6	55		
	4	70	4	85 amps
	2	95	2	115
			1/0	150
			3/0	200

Identifying and Sizing Circuits

Every receptacle on the floor plan should be identified by circuit number. Circuit-breaker spaces in the panel will have their load assignments as shown in the table on page 23.

In this electrical system there are two general-purpose circuits present in each room except the kitchen, dining room, and laundry. This scheme adds to the cost of the system, but it also ensures that if a circuit blows, there will still be some power in every room. However, you should be aware that in some localities it is common practice, sometimes even a code requirement, that lights and receptacles be put on separate circuits. Where possible, multiwire branch circuits (3-wire circuits that do the work of two 2-wire circuits) will be used. The wiring in the garage will be exposed and everything else will be concealed.

The wire used must have sufficient ampacity, or capacity, to carry the load being served. The cooktop, for instance, can draw as much as 21 and a fraction amperes. The wire for this appliance must be capable of safely carrying this much current, and the circuit breaker that's to be used must be sized to properly protect the wiring.

This house will be wired with Type NM cable (commonly referred to by the trade name Romex®), which is an assembly of two or more insulated conductors and one bare ground wire enclosed in a jacket of moisture-resistant, flame-retardant, nonmetallic material. A mix of two-conductor and three-conductor, as well as different sizes, will be used.

Electrical wire is sized using American Wire Gauge (AWG) sizes. The lower the wire number, the greater its size and ampacity. All wires and cables are marked with the following information: the maximum rated voltage for which it is listed, a letter or letters that identify it as to type, the manufacturer's name, and the AWG size or circular mil area. No. 14 is the smallest wire size allowed for general wiring. Choose copper wire and cable for interior wiring jobs.

The following table lists ampacities for the copper wire sizes commonly used in house wiring. Note that ampacities for No. 4 and larger may be lower in the original wiring in older homes. This is because older houses were wired with 60° C and 75° C wire, the only wire available at the time. These wires have lower ampacities than the 90° C wire now in use.

Understanding Cable

Type NM, or nonmetallic-sheathed, cable is probably the most commonly used cable in residential wiring. There are other types available, each with special application features. Type NMC is also a nonmetallic cable, but unlike Type NM, which can be run in dry locations only, Type NMC can be used in moist, damp, and corrosive locations as well. This means that it can be fished into voids in outside walls of masonry block or tile and installed in shallow chases in masonry and covered with plaster.

Type UF, underground feeder, cable can be buried directly in the earth but cannot be installed in poured concrete or plastered over. It can also be used in any dry, wet, or corrosive location.

Service-entrance cable is available in larger wire sizes, such as Type USE, for use in underground services, and Type SE, which is generally used to supply power to large ranges and clothes dryers.

Armored cable, commonly known as BX, is a fabricated assembly of insulated conductors in a flexible metal jacket that is intended for use in dry locations. Once a very popular material in house wiring, it has largely been replaced by nonmetallic-sheathed cable as a general-purpose wiring material. However, it is sometimes used to connect fixed appliances, such as food disposers, to the branch circuit wiring. It requires boxes with special clamps that keep the bushings in place.

All of these various cables, except for the service-entrance types, are available in two- or three-conductor configurations.

Determining the sizes and quantities of wire to buy must be done systematically. Most of the wire used in the example is No. 14 and No. 12. Plot each cable run on the floor plan with two or three slash marks, one slash to represent each conductor in the cable. Then scale the distances between boxes for all the two-conductor No. 14 (14-2) wire being used and add appropriate amounts for the up and down runs. Add these together along with any 14-2 home runs (final runs to the service panel). The result is a fairly accurate estimate of the amount of two-conductor No. 14 wire needed

Sizes and Types of Wire and Cable

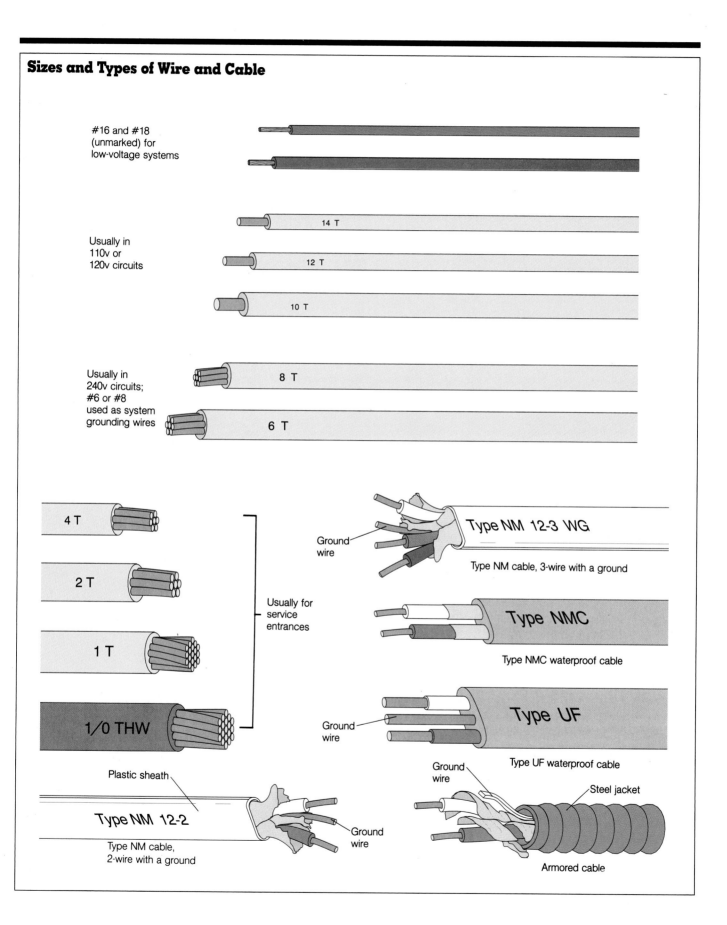

#16 and #18 (unmarked) for low-voltage systems

Usually in 110v or 120v circuits

14 T

12 T

10 T

Usually in 240v circuits; #6 or #8 used as system grounding wires

8 T

6 T

4 T

2 T

1 T

1/0 THW

Usually for service entrances

Plastic sheath

Type NM 12-2

Type NM cable, 2-wire with a ground

Ground wire

Ground wire

Type NM 12-3 WG

Type NM cable, 3-wire with a ground

Type NMC

Type NMC waterproof cable

Ground wire

Type UF

Type UF waterproof cable

Ground wire

Steel jacket

Armored cable

for the job. Do the same for the 14-3, 12-2, and 12-3 cable runs. These four sizes are generally bought in 250-foot coils, and can be purchased ahead of time. For smaller jobs, shorter coils are usually available. The larger sizes are generally sold by the foot and are quite expensive. Wait until actual cable runs can be measured to buy them. Techniques for running the wire are explained in the next chapter, "Rough Wiring."

Electrical Boxes

Electrical boxes are sold in a variety of shapes and sizes. Years ago metal boxes were the only type available. They're still widely used, but today there is also a wide range of plastic boxes available.

Switch boxes are used to house switches and receptacles. Some metal boxes have removable sides and can be "ganged," that is, mounted side by side to accommodate two or more switches or receptacles. Plastic boxes are not gangable, but they are manufactured in one-gang, two-gang, three-gang, or four-gang styles. All of these boxes are available in a variety of depths. Nailing is a common way of mounting these boxes in new work.

Round metal boxes and octagonal metal boxes are used most often with ceiling fixtures. Some employ integral mounting brackets and nail up, while others mount on a bar hanger that is attached to the house framing. One special type of ceiling box is listed for use with ceiling fans.

Metal and plastic square boxes are often used as junction boxes when surface-mounted.

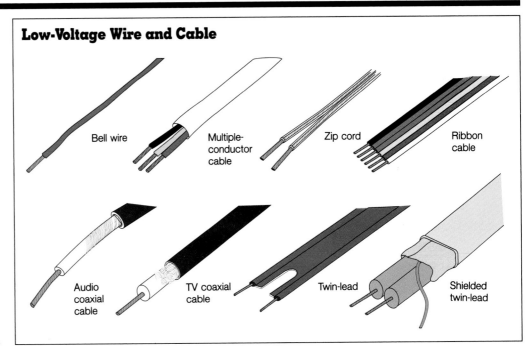

Low-Voltage Wire and Cable

Bell wire

Multiple-conductor cable

Zip cord

Ribbon cable

Audio coaxial cable

TV coaxial cable

Twin-lead

Shielded twin-lead

Choosing the Right Box

As evident in the illustration on the opposite page, the variety of boxes available is staggering. The two main categories are boxes for new construction and cut-in boxes for remodeling work.

Switch Boxes (Utility Boxes, Handy Boxes)

These are rectangular boxes that hold switches and receptacles. The basic size is 2 inches by 3 inches, which means the rectangular opening is not much larger than the switch or receptacle to be mounted in it. Some metal switch boxes have removable sides so they can be ganged together to make larger boxes. Switch boxes are sometimes used where wall bracket lights are installed and for telephone outlets.

Ceiling or Outlet Boxes

These octagonal or round boxes are designed to be covered with a light fixture or porcelain lamp holder that has essentially the same shape. They also are sometimes used to hold receptacles.

Box Extension

This is a backless box that can be screwed onto another box, piggyback fashion, to make a deeper box.

Square or Junction Boxes

Square boxes serve a variety of needs. They are used with mud rings to flush-mount switches and receptacles where large box capacity is needed. When surface-mounted, they are used as junction boxes and covered with a blank cover, or used with raised covers that hold switches or receptacles.

Special Boxes

These boxes are designed to meet specialized needs. One such box is intended for use with ceiling fans. It is designed and installed so the framing, not the box, supports the weight of the fan. Another is the ceiling pan. Only ½ inch deep, it can be mounted directly on the ceiling joist or on the surface of the ceiling. Either way, a single cable enters through the back of the box, and the box is completely covered by the fixture attached to it. Shallow switch boxes, slightly over 1 inch deep, are made for use on concrete-block walls where 1-inch furring strips support the finished wall. Special waterproof boxes are made for use outdoors.

Types of Boxes

Nail-on nonmetallic handy box

Cut-in nonmetallic box

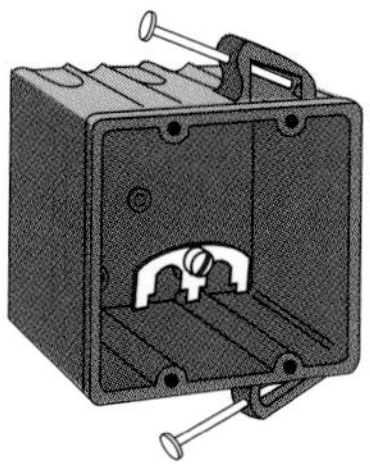

Nail-on nonmetallic 2-gang box

2-gang nonmetallic box with side brackets

2-gang nonmetallic box with front brackets

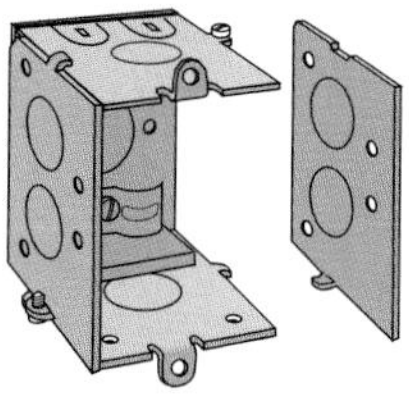

Square-corner metal switch box

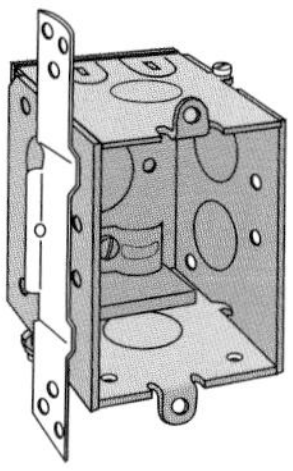

Square-corner metal switch box with side nailing bracket

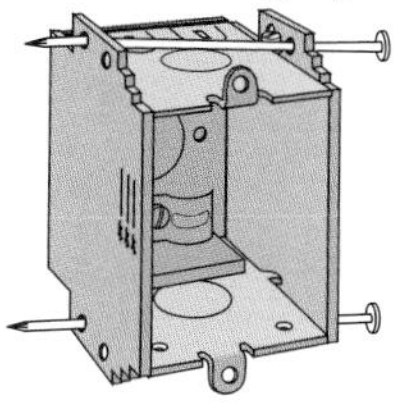

Square-corner metal switch box with 16d nails in side

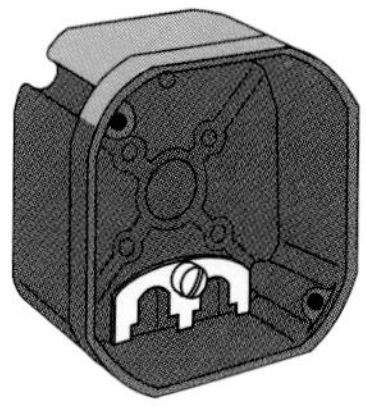

4-inch nonmetallic ceiling box

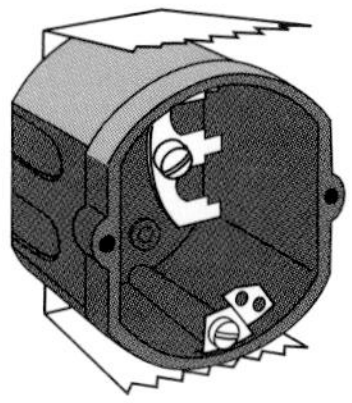

3.5" nonmetallic ceiling box

Square metal junction box

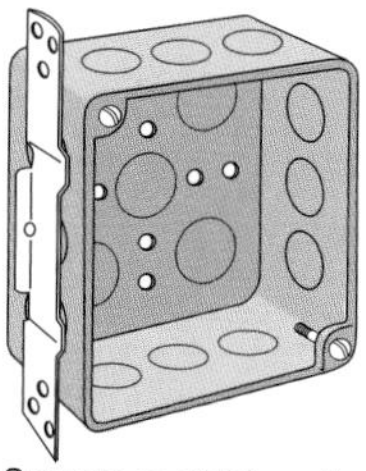

Square metal junction box with side bracket

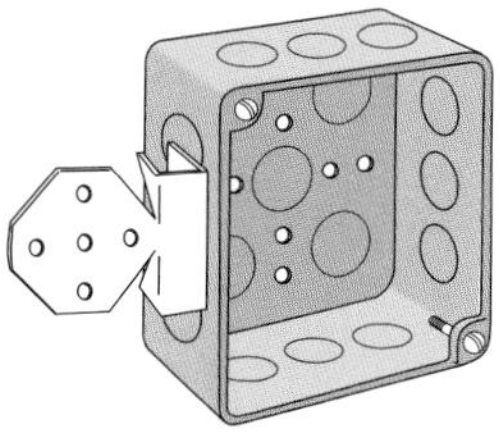

Square metal junction box with front bracket

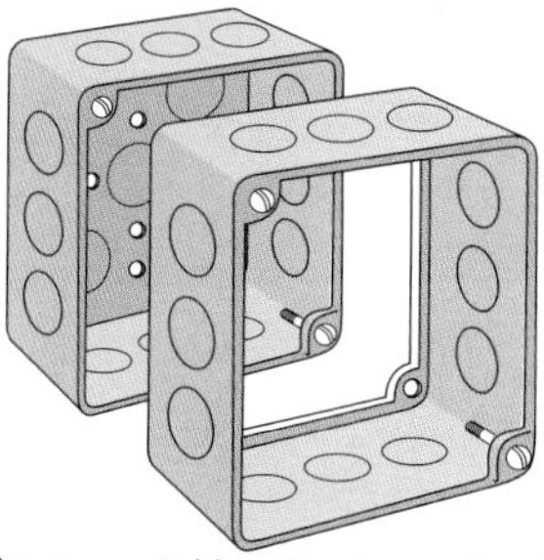

Square metal junction box extension

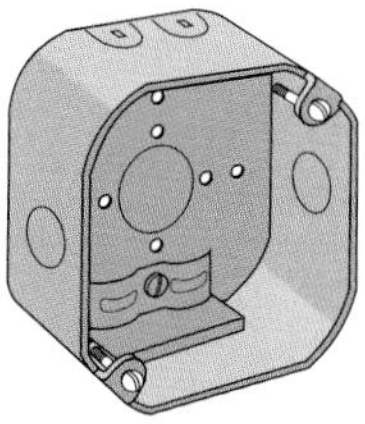

Octagon metal junction box

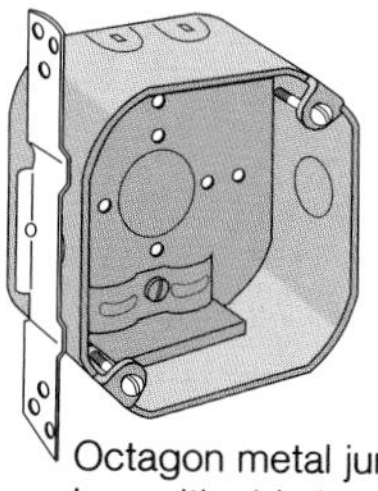

Octagon metal junction box with side bracket

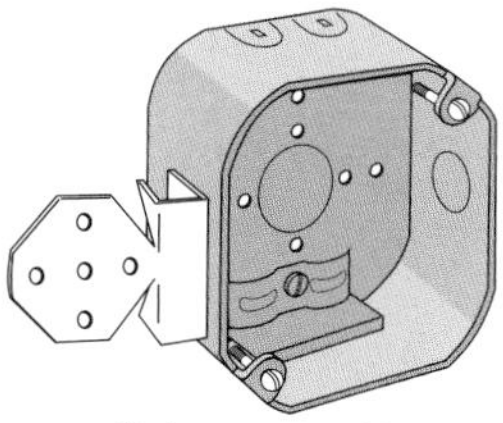

Octagon metal junction box with front bracket

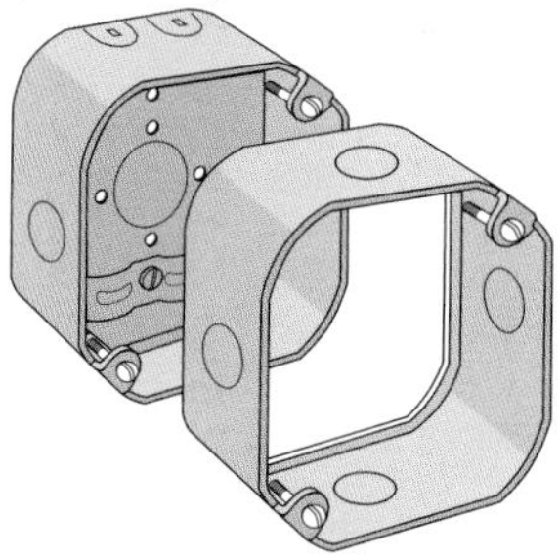

Octagon metal junction box extension

Square box flat cover

Square box flat device cover

Square box raised device cover

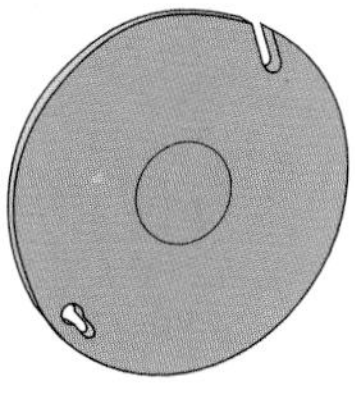

Round box cover

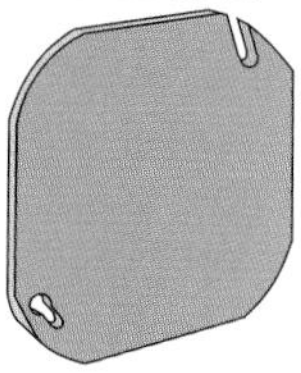

Octagon box cover

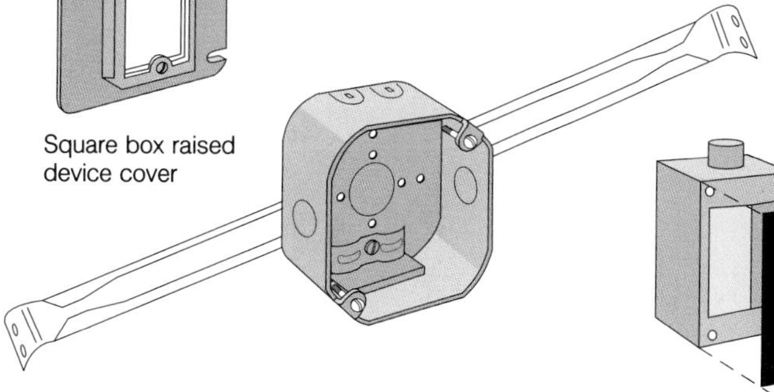

Stud bar hanger for ceiling boxes

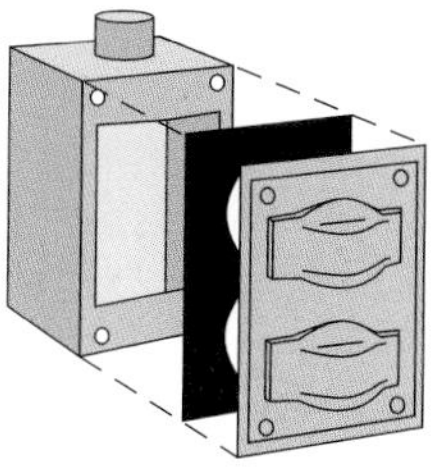

Weathertight box

Reading a Receptacle

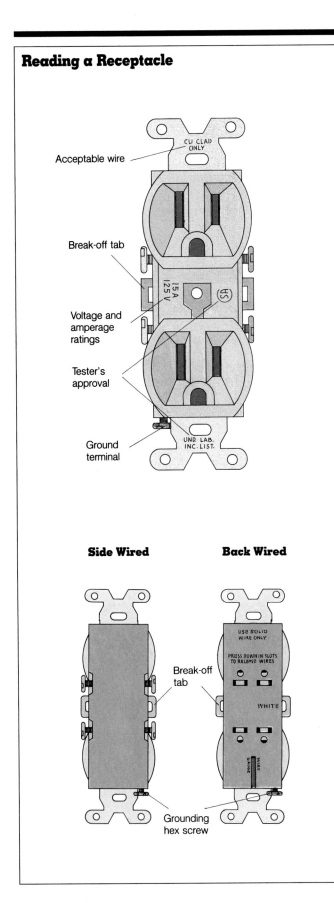

Acceptable wire

Break-off tab

Voltage and amperage ratings

Tester's approval

Ground terminal

CU CLAD ONLY

15 A 125 V

UND. LAB. INC. LIST.

Side Wired

Back Wired

Break-off tab

USE SOLID WIRE ONLY

PRESS DOWN IN SLOTS TO RELEASE WIRES

WHITE

WIRE GAUGE

Grounding hex screw

GFCI Devices

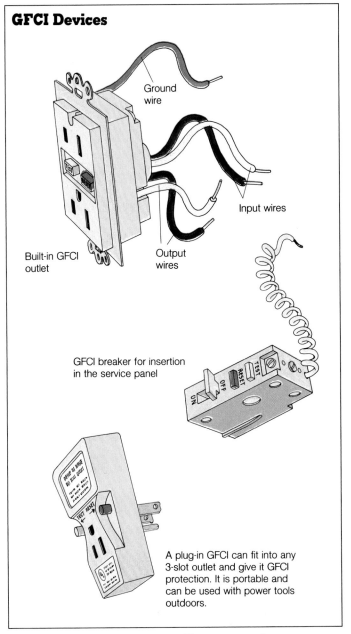

Ground wire

Input wires

Built-in GFCI outlet

Output wires

GFCI breaker for insertion in the service panel

RESET TEST OFF ON

A plug-in GFCI can fit into any 3-slot outlet and give it GFCI protection. It is portable and can be used with power tools outdoors.

When square boxes are used to hold switches and receptacles in concealed work, they are combined with a flat cover that has a raised, switch box–shaped opening that is called a mud ring or plaster ring. Handy boxes are slightly longer than switch boxes and are intended for surface mounting a single switch or receptacle. A number of other boxes are available, mostly for special applications.

Receptacles

Receptacles are used to connect equipment with a cord and plug to the electrical system. Grounding receptacles are used in all new work. The most common type 120-volt receptacle is the duplex, one that has

Reading a Switch

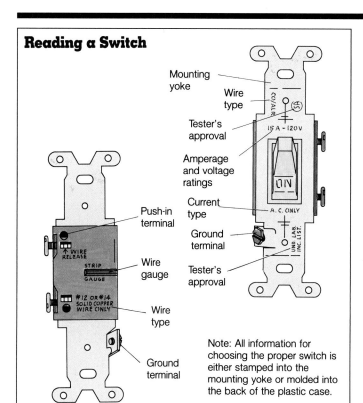

Mounting yoke

Wire type

Tester's approval

Amperage and voltage ratings

Current type

Ground terminal

Tester's approval

Push-in terminal

Wire gauge

Wire type

Ground terminal

Note: All information for choosing the proper switch is either stamped into the mounting yoke or molded into the back of the plastic case.

spaces for two plugs. Power is connected to its two parallel slots, and its U-shaped opening is connected to ground.

Ground fault circuit interrupters (GFCIs) provide greater protection than ordinary grounding-type receptacles. The *NEC* requires them in bathrooms, unfinished basements, garages, and along countertops within 6 feet of a sink. They're also required for most outside electrical outlets.

A 20-amp receptacle will be used on the laundry circuit; all other 120-volt receptacles will be 15 amp. The oven and cooktop will be cord-connected to properly sized surface-mounted receptacles for ease of appliance servicing. A 120-volt, 20-amp circuit will be connected directly to a service

switch on the furnace. Also, a 240-volt, 30-amp branch circuit will be run to a safety switch mounted close to the air-conditioning compressor.

Switches

An ordinary two-way switch is used to control a light from a single point. Controlling that light from two points requires a pair of three-way switches. A four-way switch is required for each additional control point. Dimmers are often used in place of single-pole and three-way switches in dining rooms and kitchens. Note that fluorescent fixtures require special, more expensive dimmers. So-called "quiet" switches and "touch" switches are available for use where desired. Illuminated switches require no

Some Special Switches

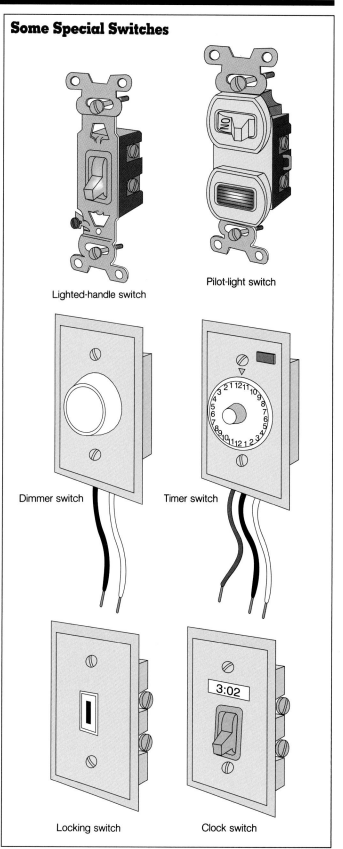

Lighted-handle switch

Pilot-light switch

Dimmer switch

Timer switch

Locking switch

Clock switch

special wiring and can be used as direct replacements for most ordinary switches. They are common in bedrooms and bathrooms.

Low-voltage relay systems are sometimes used in houses where complex light switching is desired. These systems use a separate relay to control the 120-volt power to each lighting outlet. Low-cost low-voltage switch loops are used to energize the individual relays. When a relay is energized, its light turns on. Complete systems include a master control panel, usually installed in the master bedroom, that can control all the lighting in the house from one location.

The list of electrical options available for the home is nearly endless. The most common and necessary are doorbells, telephones, and smoke detectors. As previously mentioned, smoke detectors should be installed near the furnace and on every habitable level. Exhaust fans are desirable in bathrooms and kitchens. Often an exhaust fan is combined with a ceiling light and heater in bathrooms. Other useful options include garage door openers, driveway floodlights, yard lights, underground lawn sprinkler systems, television cable, and intercoms.

Where to Buy Materials

An all-service building-supply house that's geared toward do-it-yourselfers is an excellent source of electrical supplies. Such outlets usually are well stocked and their merchandise is well labeled and backed up with good point-of-purchase information. Time spent in the electrical department can be quite an education. The best of these stores usually have experts on hand to answer questions and help with material selection.

A well-stocked hardware store is another excellent place to buy electrical supplies. Plenty of sound advice is usually available at no cost, and the staff may take a personal interest in your project as well.

Dealing with a wholesale house can be difficult if not impossible. Wholesalers sell to contractors, manufacturers, and other industrial accounts, and usually aren't interested in retail sales. They often have a minimum-purchase policy as well. Also avoid buying through catalogs because catalog descriptions often confuse even the most knowledgeable do-it-yourselfer.

How to Purchase Materials

Buy No. 14 and No. 12 wire by the 250-foot coil unless you are doing a very small project; it is far cheaper that way. Larger sizes are generally bought by the foot. Avoid buying the least-expensive switches and receptacles. Usually spending just a bit more buys much better equipment. Finally, the best rule is to buy more than you think you will need.

Obtaining a Permit

Always inquire about a permit before beginning any electrical installation. In fact, apply for the permit before buying any material. Permits are generally issued by a municipal building-permit department upon application and payment of a fee. An inspection or inspections are part of the permit process. Final approval of your installation by an inspector is a good indication that the installation is free from hazard.

Inspection schedules vary. If you have any questions about specifics, especially if a service is involved, ask the electrical inspector for clarification before you begin work. This public official is the best local resource you have and should be available to you by prior appointment.

There is a rough-in inspection that takes place when all the boxes have been installed and the wiring has been run to them. This is always done before finish materials are applied to interior walls in new work. A new service is usually inspected at this time, although the service inspection may be separate. A final inspection takes place after all connections have been made and all switches, receptacles, and so forth are in place, and branch-circuit connections in the panel are complete. It's the installer's responsibility to notify the inspection department that the installation is ready for inspection at each stage.

Plan carefully, read, and perhaps even visit a local project to see how work is being done before applying for a permit. Spend time with the inspector before starting the work, especially if a service is involved. There will be many questions about specifics, and the inspector is the best local resource there is.

Expect the inspection of the service to be stringent. This shouldn't be a problem if it has been installed employing specifics provided by the inspector (regarding grounding, bonding, connections at the terminals, and so on). The inspector will look at cable and conduit routing, protection, and support, and will check to be certain that box sizes are large enough for the number of conductors entering them, that there's adequate free conductor length in every box, and most important, that proper ground continuity is maintained throughout the system. Also, the inspector will want to see that the entire installation has been done in a workmanlike manner. This simply means he'll check to see that your work is done neatly, as a journeyman electrician would do it, with no twisted cable, disorderly or oversized loops, or boxes out of plumb, for example. Read the chapter on Rough Wiring for some helpful tips on passing inspection.

Using a Voltage Tester

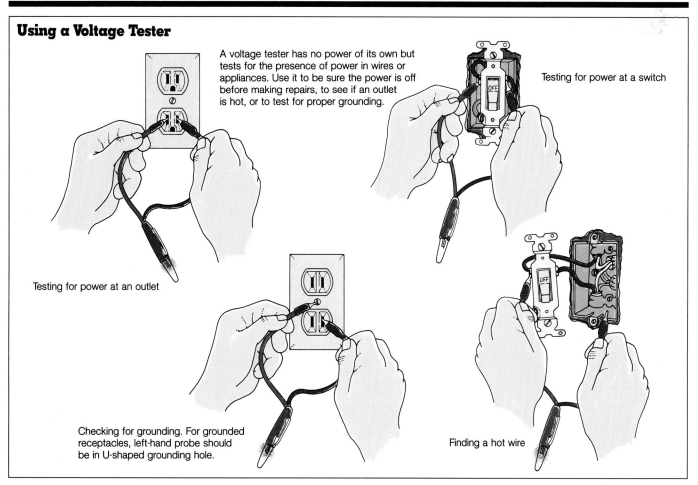

A voltage tester has no power of its own but tests for the presence of power in wires or appliances. Use it to be sure the power is off before making repairs, to see if an outlet is hot, or to test for proper grounding.

Testing for power at a switch

Testing for power at an outlet

Checking for grounding. For grounded receptacles, left-hand probe should be in U-shaped grounding hole.

Finding a hot wire

The Proper Tools

Home electricians need very few specialized tools to do small electrical projects. Installing an additional receptacle is probably the first do-it-yourself electrical installation project done in most homes. In homes with wallboard walls and the electrical panel in an unfinished basement, the only tools needed to install a receptacle in a first-floor wall are a pair of pliers for cutting wire and cable, a saw for making the cutout in the wall, an electric drill with a bit large enough for the cable being used, a screwdriver, hammer, knife, and ruler of some sort.

More complex jobs require additional tools, which may need to be purchased, including a voltage tester and a continuity tester; both are inexpensive. Finally, buy a cable stripper, a wire stripper, needle nose pliers, a wood folding rule (safer than a metal tape for electrical work), and a 9-inch nonmetallic torpedo level. If you are tackling a large, complex project, consider buying a low-cost amp-volt meter. The voltage tester is rugged and can be carried in a shirt pocket, but it can't indicate voltages below about 60 volts or above 250

volts, whereas a volt-amp meter can accurately measure a great range of voltages. A continuity tester can read only through low-resistance circuits and devices when the electricity is turned off. A switch is an example of a low-resistance device. Amp-volt meters allow the user to check motor windings, solenoid valve coils, transformer windings, and other high-resistance devices for continuity or ground.

If work is to go beyond a small nonmetallic-sheathed cable installation, especially if the work will be in masonry or plaster, additional tools needed will include a variety of screwdrivers, diagonal-cutting and side-cutting pliers, a pair of 9- or 10-inch pump pliers, a crimping tool for attaching terminals to wire, and a scratch awl. You will also need a hacksaw and a wallboard saw as well as a narrow-bladed, fine-toothed jab saw for cutting lath. Also include a variety of wood bits, twist drills, and masonry bits along with a variable-speed ⅜-inch electric drill. A 1-inch

Basic Wiring Tools

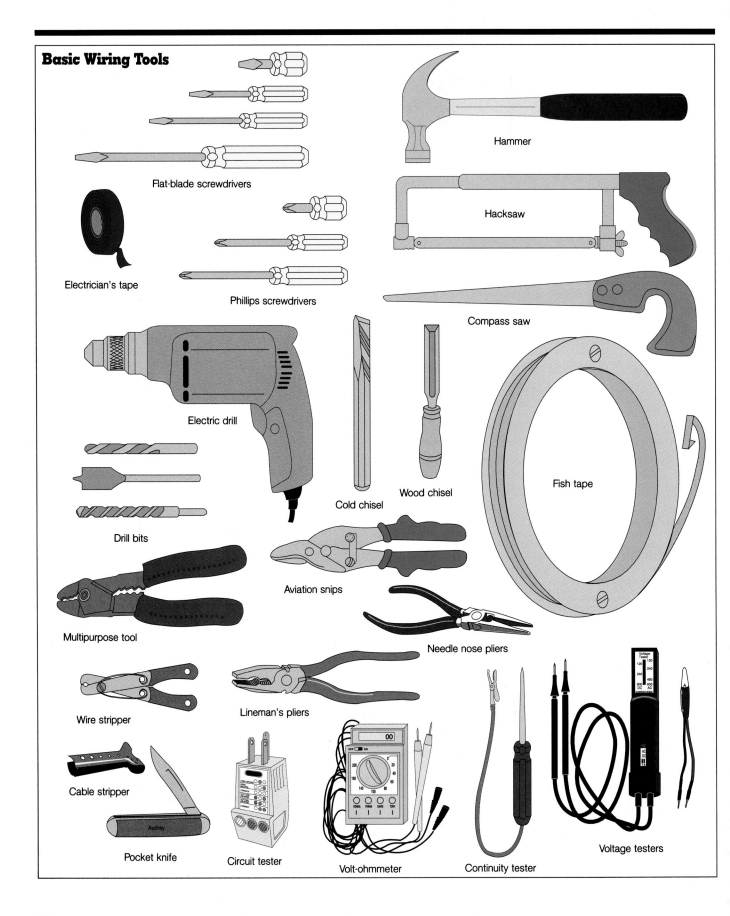

Flat-blade screwdrivers

Hammer

Electrician's tape

Phillips screwdrivers

Hacksaw

Compass saw

Electric drill

Cold chisel

Wood chisel

Fish tape

Drill bits

Aviation snips

Multipurpose tool

Needle nose pliers

Wire stripper

Lineman's pliers

Cable stripper

Pocket knife

Circuit tester

Volt-ohmmeter

Continuity tester

Voltage testers

Using a Continuity Tester

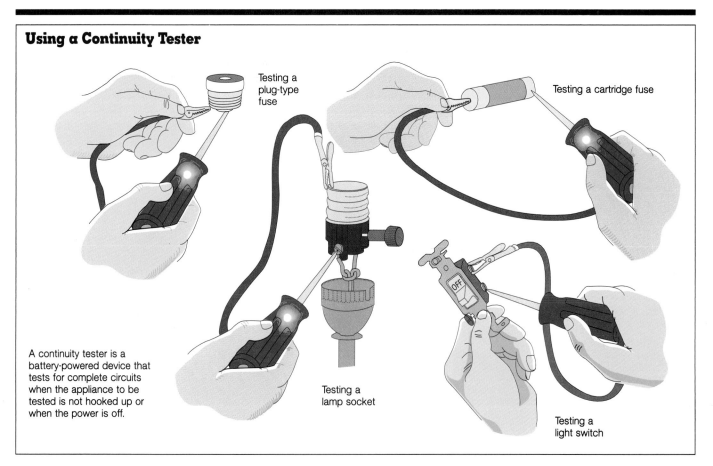

Testing a plug-type fuse

Testing a cartridge fuse

A continuity tester is a battery-powered device that tests for complete circuits when the appliance to be tested is not hooked up or when the power is off.

Testing a lamp socket

OFF

Testing a light switch

wood chisel and a 6-inch cold chisel that is ½ inch wide should also be part of the electrical tool kit.

Electronic stud finders are far superior to the older magnetic models, and they're invaluable for work in old walls. A 25- or 50-foot fish tape is another must. It's best to rent tools needed for any work requiring bending and threading conduit.

Electrical work requires planning—don't rush it. Remember that any mistakes you make can usually be corrected without too much difficulty. Start small because a big project can be overwhelming. Confidence comes with experience. Professional electricians spend up to six years in school and

on the job before they're eligible for licensing, so don't expect to learn everything you want to know about electrical wiring in a day or in a single project.

If, after much planning and rethinking, you're still not comfortable about doing the job, you should probably call an electrician. Or, if half the power in the house goes out and a look in the panel shows that a main lug has burned off—unless you have lots of time and your skill level is very high—call an electrician.

There are other situations in which it's best to call in a contractor. Some work requires special equipment, for instance, pushing conduit beneath a 20-foot concrete drive to provide

power for outdoor lighting, or bending and threading conduit for a large service. An electrician can also put his experience to work for you when a problem arises that just won't go away. Strange things do happen, such as some bulbs in the house burning brightly while others are dim, or a clothes dryer that checks out as sound but continues to blow fuses. When faced with such puzzling circumstances, you need the help of a professional.

Always keep safety uppermost in your mind when you are doing electrical work. Turning off the power protects you from electrical shock, but you should be alert to other possible hazards and work carefully. Use only properly grounded or double-insulated power tools

and three-wire cords with plugs and bodies that are not frayed or cracked.

Take care when working with ladders; if outdoors, level the feet of the ladder with pieces of wood or other rigid material and, if possible, tie the top of the ladder to the building it's leaning against. While aluminum ladders are lightweight and relatively inexpensive, they have the significant drawback of conducting electricity.

Remember that the tools being used to cut and drill material can cut and drill you, too. Always wear safety glasses or goggles when working with power tools.

ROUGH WIRING

You should now be ready to start installing the electrical system you carefully planned in the previous chapter. This process begins with the rough in, as the first stage in wiring a new house or addition is called, and it involves mounting the outlet boxes, running the cable or conduit and wires, and locating the service equipment in the stud walls and ceiling joists before they are covered with finishing materials. A knowledge of how and why this work is done is helpful even for those who won't actually perform these tasks because it provides a basic understanding of the physical structure underlying an electrical system.

The electrical service can be installed anytime after a house or addition is framed up and closed in without creating problems. The electrical rough in could also begin at this time, but it is customary to wait until after the plumbing and heating have been roughed-in to make these basic electrical installations. After the electrical, plumbing, and heating rough ins have passed inspection, the walls are insulated and finished.

Running cable and setting boxes can quickly become tedious if you are doing a large wiring job, but it's imperative that you be methodical and careful and review all your work before the walls are closed.

SETTING BOXES

The NEC requires that all wiring in the home be enclosed in a box wherever it connects to a receptacle, switch, or light fixture or where splices are made. So you start the rough in by mounting the boxes for these receptacles (shown in your floor plan) on the exposed studs and joists.

Marking Box Locations

Few hard-and-fast rules govern the setting of boxes, but they should be mounted at the same height throughout the house or addition. Standard heights are to the top of the box, 16 inches up from the floor for receptacles and 48 inches for wall switches. Boxes for wall bracket lights are usually about 6 feet above the floor. Using your electrical plan as a guide, mark the side of each stud to which a box will be mounted, at the correct mounting height. To speed up this routine job, use a story pole as a marking gauge: Take a 48-inch piece of 1 by 2 and mark 16 inches up from both ends with a horizontal arrowhead, with the point at 16 inches. You can now position the stick and mark the studs quickly and repetitively.

Switch boxes installed beside doorways must be far enough away from the opening to allow for some space between the door casing and the switch plate. If necessary, you can use blocking to gain the required space.

The face of most boxes should be flush with the finished wall. Set switch boxes and ceiling boxes so they extend into the room beyond the stud or joist a distance equal to the thickness of the finished wall or ceiling. Square boxes are the one exception to this rule. They should be mounted with their fronts flush with the framing. When the mud ring is attached, its opening will be flush with the finished wall.

Many boxes nail up to the sides of the studs with large captive nails that do not pass through the interior of the box. Instead, the nails pass through extensions at the top and bottom of the box. This system provides a solid installation. Other boxes rely on integral

Roughing-In Cable and Boxes

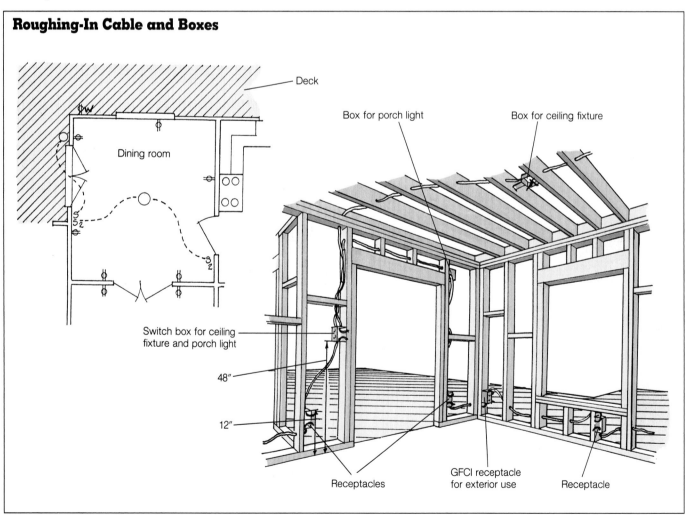

Deck

Dining room

Box for porch light

Box for ceiling fixture

Switch box for ceiling
fixture and porch light

48"

12"

Receptacles

GFCI receptacle
for exterior use

Receptacle

Common Measurements for Outlets and Switches

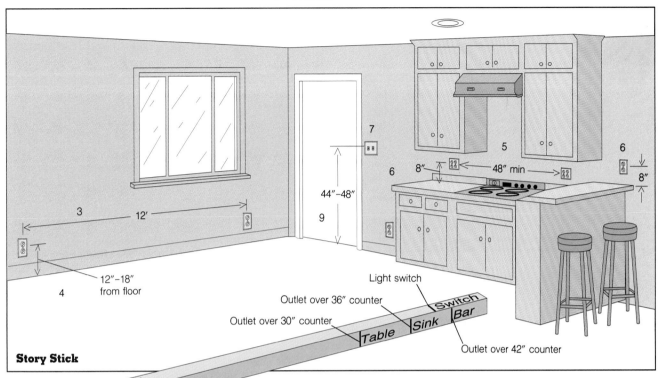

7

44″–48″

9

5

6

8″

48″ min

6

6

8″

Light switch

Outlet over 36″ counter

Switch

Outlet over 30″ counter

Bar

Table

Sink

Outlet over 42″ counter

3

12′

4

12″–18″
from floor

Story Stick

1″ × 1″ × 5′

Outlet

Floor

Floor

Center of floor outlet

Set this end
on the floor

Studs

Switch

Mark the position of the switch or outlet
box on the stud to which it will be nailed.

1. All outlets must be grounded.

2. Careful planning ensures that all receptacles on 1 floor or in 1 room are not connected to 1 circuit. In a kitchen or work area, the outlets above the counter must alternate between 2 small-appliance circuits so that adjacent receptacles are not on the same circuit. The Code allows up to 8 outlets on a circuit, but local regulations vary—be sure to check the code in your area.

3. Any wall 4′ or more wide must have at least 1 floor outlet. Any wall wider than 12′ must have at least 2 outlets. Floor outlets should be no more than 12′ apart.

4. Floor outlets should be between 12″ and 18″ above the floor.

5. Any work-counter area up to 4′ wide must have a receptacle above it; over 4′ wide must have 2 outlets. Work-counter outlets should be no more than 48″ apart.

6. Work-counter outlets should be 8″ above the counter.

7. Light switches must always be on the same side of the door as the doorknob.

8. Lights in stairways, hallways, garages, and most rooms with 2 entrances should be controlled from at least 2 locations using 3-way or 4-way switches.

9. Light switches should be between 44″ and 48″ above the floor or 8″ above a work counter.

metal brackets for mounting. The brackets are set back from the front of the box enough to allow for the thickness of common wall materials, such as wallboard and paneling, so you can set the front bracket flush with the front of the stud without measuring. Drive a couple of shingle nails through each bracket for added support.

Choosing Boxes

Boxes come in a number of shapes and sizes to accommodate particular functions and permit a specific number of devices and wires to be enclosed in them. Metal and plastic boxes are available, and to a certain extent they are interchangeable because they and the switches and receptacles used in them are standardized. The shape of a box suggests its function. (Note that in this book the term *plastic* includes a variety of nonmetallic materials such as Bakelite, vinyl, and plastic.)

Based on the sidebar on page 26 and the illustration on page 27, select the appropriate box type for each purpose and location in the house. After you have determined which shapes you need, you must choose sizes that are large enough to accommodate the number of wires and devices to be located inside them. The rules for making this selection changed with the 1990 *NEC*. Generally, larger boxes are now required because of concern about providing sufficient space to accommodate GFCIs and dimmer switches. The sidebar on page 40 shows the maximum number of wires

Extending a Circuit

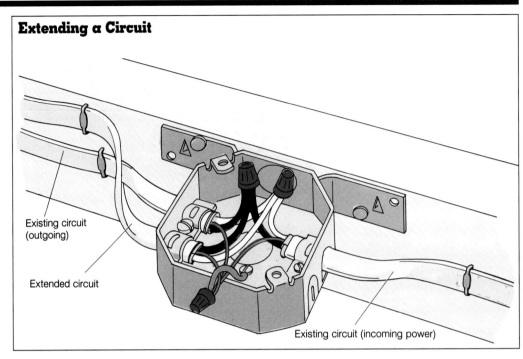

Existing circuit (outgoing)

Extended circuit

Existing circuit (incoming power)

permitted in a box according to wire size and box volume, which is stated in cubic inches. All boxes, mud rings, and surface-mount device covers are marked with their volume in cubic inches to help you calculate the permissible number of wires quickly. Here's how it works.

Each wire entering a box, except for the grounding conductors, requires space within the box based on its size. Each No. 14 wire, for example, requires 2 cubic inches of free space within the box; each No. 12 wire needs 2.25 cubic inches; each No. 10 wire, 2.5 cubic inches. The sum of these volumes largely determines the box size needed, but other factors also apply. Additional space equal to one wire must be provided for grounding conductors if one or more are present in the box, and space

equal to one wire is added for each internal cable clamp, fixture stud, and hickey. Each mounting yoke or strap containing one or more switches or receptacles requires space equivalent to two additional wires. To determine the volume required for each of these additional items, use the largest wire size in the box. Include everything to determine total volume available or required.

To summarize, your choice of box will depend on several factors, including: how the box will be used (junction box, light fixture, switch, receptacle, and so forth); how it will be mounted (against a stud, between joists, on a wall surface, in a tight space); how many switches or receptacles it will have (double gang, and so forth); how much volume is required for the size and number of wires it has; and any special requirements, such as supporting a ceiling fan or outdoor installation.

Metal Versus Plastic Boxes

Traditionally, electrical boxes were made of galvanized steel. Purists still prefer the metal box, but plastic boxes have largely replaced metal boxes in new residential construction. Both types of boxes have advantages.

Metal boxes must be used with metal conduit and cable systems. They may also be used with nonmetallic cable systems. Either way, they must be grounded. Grounding is achieved via the conduit or metal cable jacket, or by the grounding conductor in the nonmetallic cable. Metal boxes can be ganged together in various combinations. They are strong and durable, but cost two to three times more than plastic.

Installing Boxes

Metal Boxes

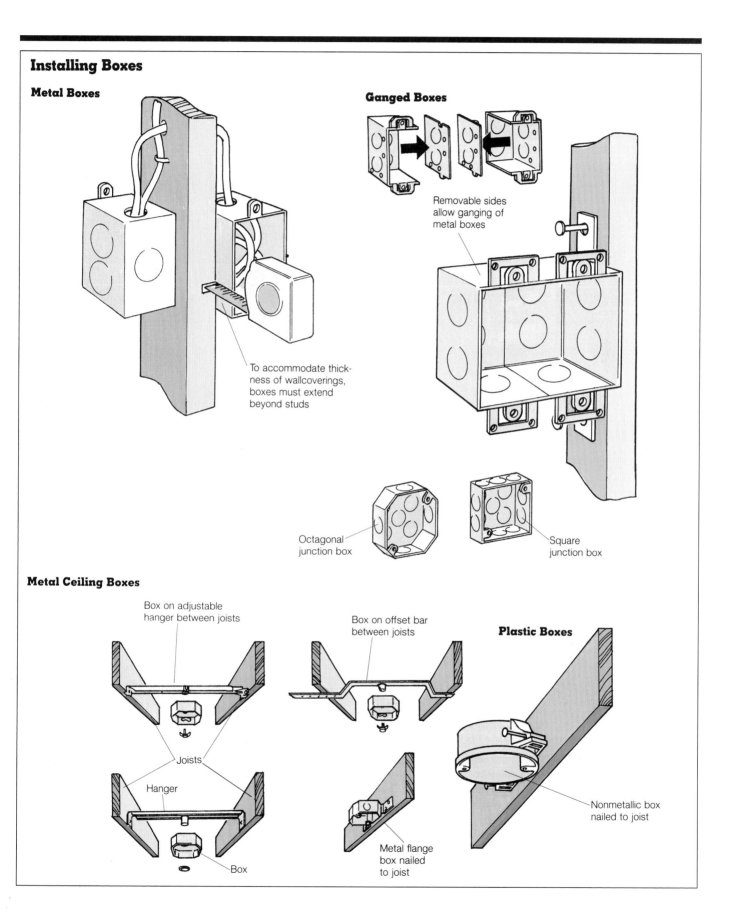

To accommodate thickness of wallcoverings, boxes must extend beyond studs

Ganged Boxes

Removable sides allow ganging of metal boxes

Octagonal junction box

Square junction box

Metal Ceiling Boxes

Box on adjustable hanger between joists

Box on offset bar between joists

Plastic Boxes

Joists

Hanger

Box

Metal flange box nailed to joist

Nonmetallic box nailed to joist

Number of Conductors Permitted in a Box

The *NEC* strictly governs the number of wires permitted in a box in order to prevent overcrowding. The number of conductors allowed is determined by two things: the size or volume of the box and the size of the wire. This table lists the number of No. 14 and No. 12 wires that can be installed in some of the most commonly used sizes of plastic boxes. The table assumes the box has grounding wires and internal cable clamps. It also assumes the one-gang boxes contain one strap-mounted device, the two-gang contain two devices, etc. You can add one wire if there aren't cable clamps in the one-gang boxes only.

Box	Conductors
13.5 cu. in. 1-g switch box	2 - #14 or 2 - #12
16 cu. in. 1-g switch box	4 - #14 or 3 - #12
18 cu. in. 1-g switch box	5 - #14 or 4 - #12
32.5 cu. in. 2-g switch box	10 - #14 or 8 - #12
46 cu. in. 3-g switch box	15 - #14 or 12 - #12
60 cu. in. 4-g switch box	20 - #14 or 16 - #12
22.5 cu. in. ceiling box (no device)	9 - #14 or 8 - #12
36.6 cu. in. 1-g square box	14 - #14 or 12 - #12
39.1 cu. in. 2-g square box	13 - #14 or 11 - #12

Plastic boxes are common in new construction because they are less expensive. They also speed up the rough in because in some cases they do not have to be grounded and can be used without cable clamps if the cables entering them are properly supported outside the box. However, they cannot be used with conduit.

Boxes for use with recessed lighting fixtures are usually a part of the fixture assembly. There are two reasons for this arrangement. First, it provides secure mounting for the box in a location that permits access to the wiring when the fixture is removed from its mounting frame. Second, because the box is not in physical contact with the fixture, except for the high-temperature wiring between them, the house wiring is not subjected to the heat generated by the fixture. This is an important safety feature.

Mapping Your Existing Circuits

Before planning a major remodeling job or addition, you should get acquainted with your existing electrical service. In other words, know what you have to work with. If there is no floor plan that documents the electrical layout of the house, this is a good time to make one. It is one of the most useful documents you can have in your home-management file.

First, go to the service panel and label every circuit breaker or fuse with a number if they are not already labeled. Next, draw the floor plan of the house. This drawing doesn't need to be meticulous or to precise scale but it should be legible, and there should be a separate sketch for each floor, including the basement and attic. Now, using the electrical symbols shown on page 20, plot the approximate location of all the switches, receptacles, and permanent lighting outlets on each floor, and then draw dashed lines between the switches and the receptacles and light fixtures they control.

Also draw in the outlets for the 240-volt and 120-volt individual branch circuits and the 120-volt small-appliance circuits. Now begin tracing the circuits in the house.

Choose a time when the house is quiet. Plug a radio into a receptacle and turn it on. Go to the panel and flip breakers off and on, or unscrew fuses, until the radio goes off. Mark that receptacle on the drawing with the corresponding circuit number. Return to the radio, move it to another receptacle, and repeat the process.

This won't be as time-consuming as you might think because you soon will notice patterns emerging, and you will be able to guess with a fair degree of accuracy the circuit to which the next receptacle is connected. You may find, for example, that all the lighting receptacles are on separate circuits, although it is more likely they will be on one of the general circuits in the room in which the lighting is installed. If you have someone to work with you, moving the radio while you flip breakers and record circuit numbers, the job will go quickly.

Once you have traced the general-purpose branch circuits, go to the kitchen area and trace all the small-appliance circuits using the radio as your guide. Identify the 120-volt individual branch circuits by tripping the remaining circuit breakers, one at a time, and checking the dishwasher, food disposer, and so forth, to see if they run. Label these receptacles with the appropriate circuit number. Use this same method to identify the 240-volt individual branch circuits as well.

Once completed, this drawing can be used to determine at a glance which outlets are on which circuits. It also reflects the branch-circuit loading throughout the house. You certainly don't want to extend a branch circuit that is already heavily loaded. It's also helpful to know in advance everything that is going to be turned off when a given circuit breaker is opened.

ROUGH WIRING FOR NEW CONSTRUCTION

It's a common practice to plan runs from the panel to a few boxes located in the center of areas where there are to be a number of receptacles, and then to wire from box to box in that area. These runs to the panel are called home runs, and they are shown on electrical drawings as short arrows connected to the home-run boxes and pointed toward the panel.

Drilling Holes

Holes for the cable runs are drilled in the framing after all the boxes have been mounted. They should be bored in the center of the stud so their edges are not less than 1¼ inches from the edge of the framing. If this clearance cannot be maintained, or if it is necessary to notch studs, the cable must be protected from nails by a steel plate that is at least 1/16 inch thick and large enough to cover the wiring area. These plates are called nail plates, and should be available wherever wiring supplies are sold.

Drill holes for cable runs that will connect wall receptacles about 1 foot above the boxes. Use drill bits that are large enough to accommodate the wire size to be used. You can use a hole for more than one cable if the hole is large enough. Usually a ¾-inch hole is large enough for one or two cables to be pulled through it. As you work, keep the holes aligned to make it easier to pull cable through them later. This requires conscious effort because the locations for holes usually aren't marked on the framing. Snap a chalk line across the bottoms of the ceiling joists or the front of the

wall studs to provide a guide, then work backward toward your destination while keeping an eye on the first couple of holes you drilled. Remember to always wear safety goggles and head protection when drilling overhead.

A ship auger is superior to most other drill bits for rough-in work. This bit is designed to pull itself through wood, letting the drill motor do the work. Most other types of drill bits require a lot of pressure to bore through the wood. This is an important consideration when there is a lot of drilling to be done, especially in tight quarters. Avoid nails; they ruin drill bits.

Pulling NM Cable

Before you start pulling cable, familiarize yourself with the rules listed below that govern the securing and stapling of cable. These rules are important. They determine what supplies you must have on hand before you begin the work, as well as how you will do that work.

Pull the long cable runs first; use any leftover short lengths of cable for the runs between the boxes. Don't permit the cable to sag excessively between holes, but don't pull it too tight, either. Open new boxes of cable as they are needed. Lay the cable alongside the run and cut it so there will be 8 or 10 inches of free cable inside each switch box. Use a cable ripper to remove enough jacket so that about ½ inch of the jacket will show inside the box, or beyond the internal cable clamps if they are used. The electrical inspector will probably want the cable ends left outside the panel until after the rough-in inspection. Leave them long enough to easily reach any point inside the panel, and identify them by circuit number.

Where cable types NM, NMC, and UF are used in exposed work (in garages and unfinished basements, for example), they should be run along the center of the sides of the joists, or through bored holes, or on running boards nailed across the joists, providing a surface for cable to be fastened to. Take care not to bend the cable too tightly. The minimum radius bend allowed is five times the diameter of the cable.

The *NEC* requires that cable types NM, NMC, and UF be secured at least every 4½ feet wherever they are not run through bored holes, and within 12 inches of every box or panel where clamps are used to secure the cable. Clamps

aren't required in plastic one-gang switch boxes if the cable is secured within 8 inches of the box.

Select and use the correct staple sizes for the cables being installed. Application charts are usually printed on the staple package. Both bare and insulated staples are available. Choose whichever is used locally. Take care not to drive staples into the cable jacket.

Special Runs

Some special rules govern the installation of wiring in attics, crawl spaces, and garages. Where cable is run across the top of floor joists in attics that are accessible by permanent stairs, the cable must be protected on both sides by guard strips at least as high as the cable. Where only a scuttle hole leads to an attic, this protection is required only within 6 feet of the entrance. In garages and basements where cables are not run inside walls, they must have physical protection wherever they are exposed within 8 feet of the floor. The most common form of protection is thin-wall conduit. A thin-wall connector and bushing are needed on the upper end of the conduit. The conduit is attached to the box using a standard thin-wall connector, which also serves to ground the conduit. Wiring in crawl spaces sometimes requires special mechanical protection, too. Consult the local electrical code.

Running Cable

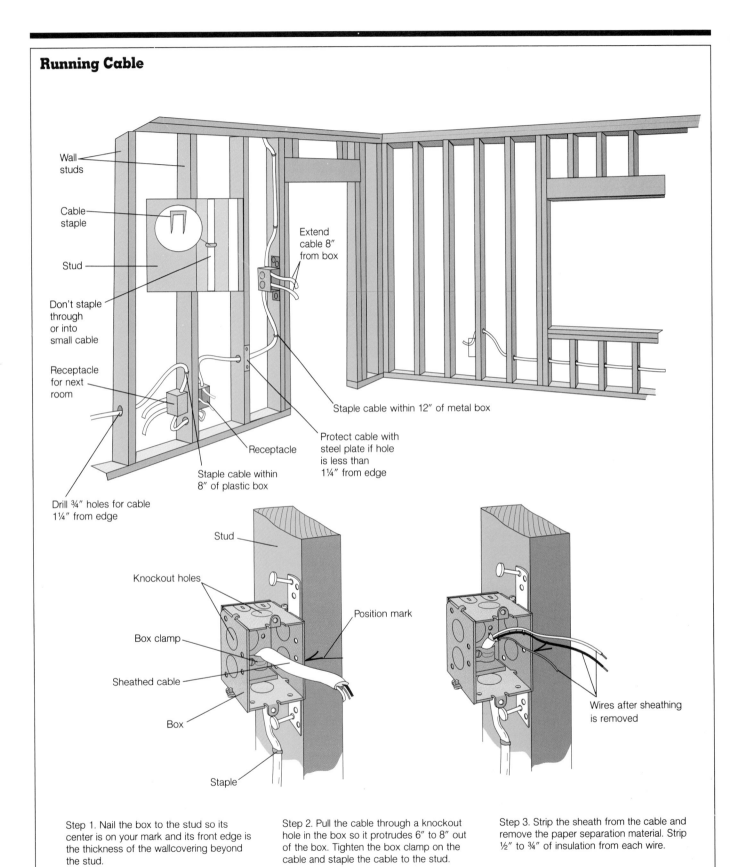

Wall studs

Cable staple

Stud

Don't staple through or into small cable

Receptacle for next room

Drill ¾" holes for cable 1¼" from edge

Extend cable 8" from box

Staple cable within 12" of metal box

Protect cable with steel plate if hole is less than 1¼" from edge

Staple cable within 8" of plastic box

Receptacle

Stud

Knockout holes

Box clamp

Sheathed cable

Box

Staple

Position mark

Wires after sheathing is removed

Step 1. Nail the box to the stud so its center is on your mark and its front edge is the thickness of the wallcovering beyond the stud.

Step 2. Pull the cable through a knockout hole in the box so it protrudes 6" to 8" out of the box. Tighten the box clamp on the cable and staple the cable to the stud.

Step 3. Strip the sheath from the cable and remove the paper separation material. Strip ½" to ¾" of insulation from each wire.

Using BX Cable

Years ago, armored cable, commonly known as BX, was used extensively in residential wiring, but nonmetallic cable has largely replaced it as the wiring method of choice. However, BX cable is often used in confined spaces such as inside cabinets, cupboards, and for runs to range hoods and food dispersers, because it is superior mechanically to nonmetallic cable.

Actually, BX can be used in any dry place above the bottom of the basement ceiling joists. It can't be used in damp or wet locations because paper spacers are used to maintain separation of conductors inside the jacket, and the paper is vulnerable to moisture. Red antishort bushings are inserted into the cable ends to protect the conductors from the edge of the metal jacket; and special connectors, which are designed to keep the bushings in place, are used to connect the cable to the boxes.

Wiring Smoke Detectors

Smoke detectors have become standard equipment in homes. Originally, all smoke detectors were battery powered, and that type is still most commonly found in older homes. Information about types and location requirements is available through your local building-inspection department. Ideally, a house should contain a mix of battery-powered and hard-wired alarms because it's unlikely that a battery and the home's electrical system would fail at the same time.

Installing Armored Cable

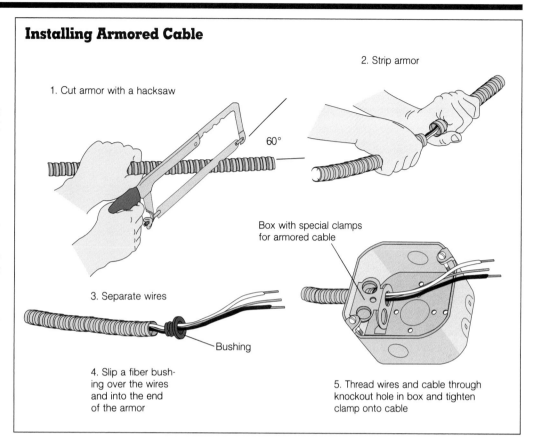

1. Cut armor with a hacksaw

60°

2. Strip armor

3. Separate wires

Bushing

4. Slip a fiber bushing over the wires and into the end of the armor

Box with special clamps for armored cable

5. Thread wires and cable through knockout hole in box and tighten clamp onto cable

BX Cable Applications

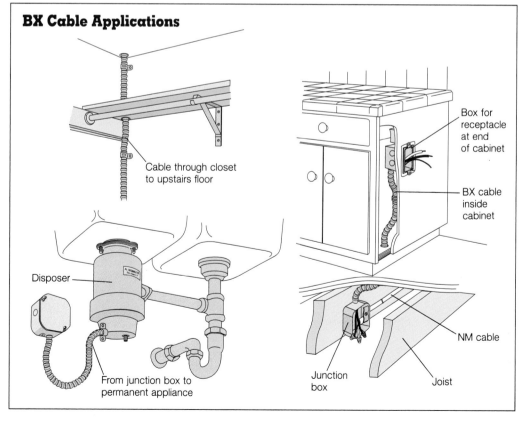

Cable through closet to upstairs floor

Disposer

From junction box to permanent appliance

Box for receptacle at end of cabinet

BX cable inside cabinet

NM cable

Junction box

Joist

When hard-wiring smoke detectors, it's advisable to put them on the same circuit as lights which are turned on several times a day so that you will know and respond quickly if the circuit goes out.

Wiring for Specialized Equipment

Doorbells and thermostats are wired with low-voltage wire, usually No. 18 or smaller (bell wire). This is usually installed after the 120/240-volt wiring is in place. The thermostat is in a circuit that is powered by a 24-volt source located in the furnace, and it should be mounted about 4 feet off the floor on an inside wall near the center of the house.

A separate transformer must be installed for the doorbells. This is a small 12-volt unit that usually mounts in a knockout in the side of a ceiling box in the basement or garage, where it's accessible for future replacement if necessary. Small holes are drilled for the cable where needed, but whenever possible the cable is run through the holes used for the main house wiring. Insulated staples are used to secure the cable where it is run along the framing of the house.

Wiring for telephone, stereo, security systems, intercoms, remote-control systems, and so on, should be installed before the walls are closed up. Detailed installation instructions are always included with packaged systems; follow such instructions exactly.

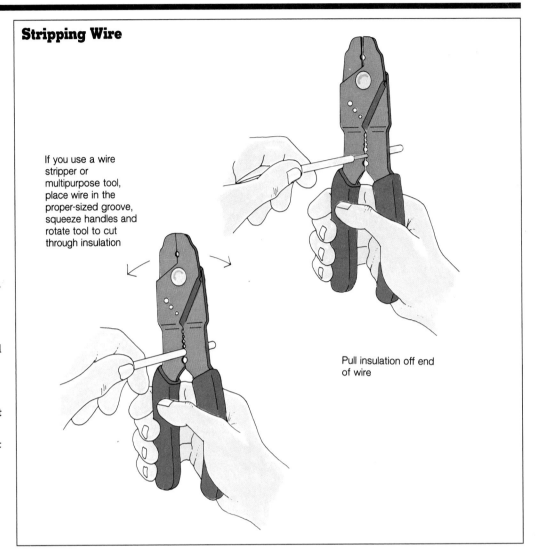

Stripping Wire

If you use a wire stripper or multipurpose tool, place wire in the proper-sized groove, squeeze handles and rotate tool to cut through insulation

Pull insulation off end of wire

Labeling and Marking Wires

Cables that terminate in the service panel should be marked with their circuit numbers or destinations using indelible felt-tipped markers. Most of the boxes in the example house have one cable entering and one cable leaving them. Connections here should be obvious, and any question about the circuit number involved can be answered by checking the electrical layout on the floor plan. Real confusion can occur at ganged boxes where more than one switch is to be mounted; to minimize this, gently twist together the wires that go to each switch. When it comes time to do the finish work, the connections will be obvious.

Splicing

All splices must be made inside an electrical box or other approved housing. Most splices in residential wiring are made using wire nuts. These wire nuts are available for wires as large as No. 6. Color is used to identify size. Two of the most frequently used sizes are yellow and red, with red being the larger of the two. Wire nuts available now are generally approved for copper-to-copper connections only. Wire-nut packages contain charts that list allowable wire combinations: Three No. 14s, two No. 12s, a No. 9 and a No. 10, etc. Additional information from the manufacturer includes strip length and whether the wires are to lay side by side or be twisted together before turning on the wire nut. Follow all of these instructions carefully.

The wire nuts used to connect circuit conductors together

Splicing Wire

Joining Wire to a Screw Terminal

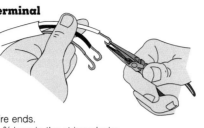

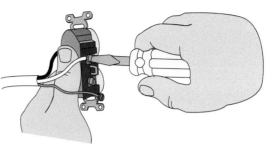

Strip ¾″ of insulation from the wire ends.
Use needle nose pliers to form a ¾-loop in the stripped wire.

Hook the bent wire clockwise around the screw terminal and tighten the screw.

Joining Wires With Wire Nuts

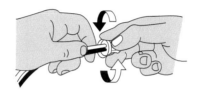

Strip about ½″ of insulation from the ends of wires you're joining. Stick the ends into the wire nut and turn it clockwise until tight.

Note: Some manufacturers of wire nuts require you to pre-twist the wires before screwing on the nut

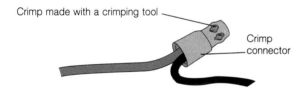

Crimp made with a crimping tool — Crimp connector

Strip insulation as you would for a wire-nut splice. Insert the wires and crimp the connector.

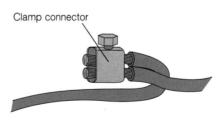

Clamshell connector — Lid — Metal insert with points

This connector is molded from one piece of plastic and has a metal insert with points. You insert the wires and close the clam shell, forcing the metal points through the insulation into the metal wire.

Joining Several Wires to a Screw Terminal

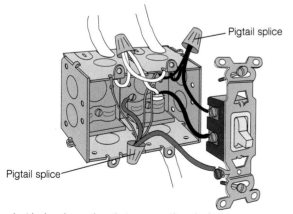

Pigtail splice

Pigtail splice

The electrical code requires that no more than 1 wire be attached to a screw terminal. To attach more wires to a terminal, splice them together along with another short piece of wire (the pigtail) with a wire nut. Then attach the pigtail to the screw terminal.

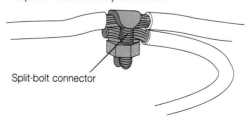

Clamp connector

Bare wires and connector must be wrapped with electrician's tape or enclosed in a junction box.

Split-bolt connector

in NM cable may also be used to connect grounding conductors together. A special wire nut is available for grounding purposes in metal boxes. It resembles an ordinary wire nut, except that it is green and has a hole through its top. A bare grounding wire passes through the hole, with one end connected to the ground screw in the box and the other end connected to the ground terminal on the switch or receptacles being wired. The ends of the ground wires from the cables entering the box are pushed into the wire nut, and it is then twisted tightly over the entire group of wires until it is tight.

Crimp-on terminals are needed for certain applications, in which an ordinary screw is used to connect a stranded wire to a terminal, for example. Without the crimp-on terminal, the stranded wire would flatten out, and some of the strands would not be covered by the screw head. Terminals are classified by wire size and screw size. A 10-10 terminal is meant to be used with No. 10 or No. 12 wire and a 10-24 or 10-32 screw. These terminals are available both bare and insulated in a number of configurations. A crimping tool, preferably one that also strips wire, is a worthwhile investment for the home electrician.

Using Conduit

It's often wise, and sometimes required, to install conduit and wire rather than cable in basements, garages, and some out-

door wiring situations if the cable will be subjected to physical damage. Conduit is used in all service installations, although this hasn't always been true. Conduit systems are continuous from box to box, which means that future wiring changes and additions can be made easily.

Conduit runs must be complete before wire is pulled into them. Wire is usually fished from outlet to outlet with a fish tape. To do this, first, push the fish tape through the conduit. Then strip about 4 inches of insulation from the end of each of the wires that will be pulled through the conduit. Push these bare ends halfway through the eye of the fish tape, fold them back 180 degrees, and tape the group closed with two or three wraps of electrical tape. All of the bare wire should be covered. Now, one person pulls the fish tape through the conduit while a helper feeds the wire into the conduit and keeps it straight. Wire usually can be pushed into short, straight conduit runs with ease if the wire ends are taped together as described above and the group is kept relatively straight while pushing.

EMT (Thin-wall) Conduit

Electrical metallic tubing, commonly called EMT, thin-wall, or steel tube, is a popular, rather easy-to-install conduit that is often used in basement and garage wiring. It's also approved for outdoor installations when used with weathertight couplings and connectors. EMT

is cut with a hacksaw using a fine-toothed blade; never use a tubing cutter. After the conduit is cut it must be reamed and deburred. Use channel lock pliers for both operations. To deburr, gently close the pliers over the cut end of the conduit and rotate the pliers back and forth while holding the EMT in your other hand. To ream, open the pliers until the tips of the handles come together, and then insert them into the conduit and twist back and forth, or if the pliers' handles are cushioned, insert the upper jaw into the conduit and twist. Either way, it's easy; the material is soft.

EMT is never threaded. Instead, it is coupled together and connected to boxes using couplings and connectors that slide ¾ inch or so over the thin-wall and are held firmly to it using set screws (indoors) or a compression arrangement (outdoors). Fittings using an indenting tool were once popular because of their low cost, and you may find this kind in your home. A locknut is threaded onto the connector to hold it securely to the box. Generally, bushings aren't required.

Because it bends easily, ½-, ¾-, and 1-inch EMT can be bent with a hand bender; 1¼-inch can be bent by hand, but is more difficult. Larger sizes require mechanical bending. Use the correct-sized bender for each EMT size. Factory-bent elbows are available in all sizes. Offset connectors can be used

when only a small amount of offset is needed where the EMT enters a box.

EMT cannot be installed with more than the equivalent of four quarter bends (360 degrees) between boxes. Bend it more than this and it may be impossible to get a fish tape through it for pulling wire. EMT must be supported with approved straps at least every 10 feet and within 3 feet of each box or panel.

Rigid Metal Conduit

This type of conduit is often used for the mast in overhead services, and is always used in low-roof services, that is, where the mast extends above the roof of the home. Rigid metal conduit is threaded and reamed on both ends. It joins together with couplings that thread over the conduit, and is fastened to boxes and panels with a locknut or locknuts (one outside and one inside the box) and a bushing on the inside to protect the wire from abrasion.

Small sizes can be bent by hand, but larger sizes should be bent mechanically. A variety of factory elbows and L- and T-shaped conduit bodies are available. The rules for supporting rigid metal conduit and the allowable number of bends between openings are the same as for EMT.

Rigid is truly the "use everywhere" conduit. It offers maximum mechanical protection of the wires within it, indoors, outdoors, and in the ground. Black enameled conduit is sometimes found in

Conduit

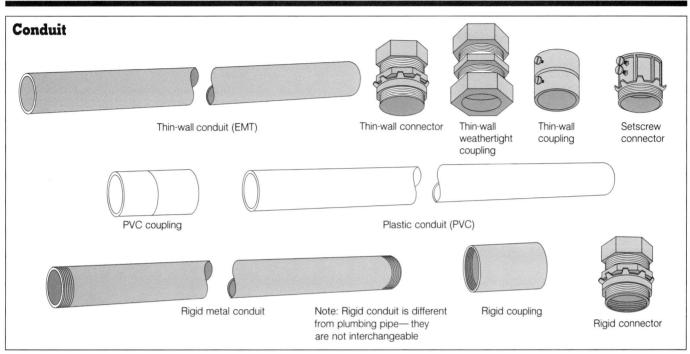

Thin-wall conduit (EMT)

Thin-wall connector

Thin-wall weathertight coupling

Thin-wall coupling

Setscrew connector

PVC coupling

Plastic conduit (PVC)

Rigid metal conduit

Note: Rigid conduit is different from plumbing pipe— they are not interchangeable

Rigid coupling

Rigid connector

Bending Conduit

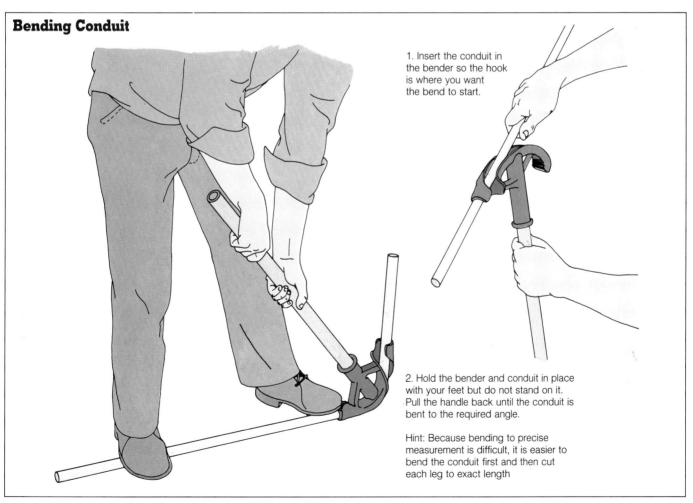

1. Insert the conduit in the bender so the hook is where you want the bend to start.

2. Hold the bender and conduit in place with your feet but do not stand on it. Pull the handle back until the conduit is bent to the required angle.

Hint: Because bending to precise measurement is difficult, it is easier to bend the conduit first and then cut each leg to exact length older homes; this is ordinary conduit that has had a coat of enamel applied to it by the manufacturer. It can be used only indoors because it is not galvanized.

PVC (Rigid Nonmetallic) Conduit

This nonmetallic conduit is a PVC product that is used primarily outdoors, especially for service-entrance conductor raceways and wiring to swimming pools and spas. Like EMT and rigid metal conduit, it is manufactured and sold in 10-foot lengths.

This rigid PVC conduit can be cut with a hacksaw, and the rough edges can easily be trimmed inside and out with a knife. Sections are joined together with couplings that are attached to the conduit ends with solvent cement. Connections at boxes and conduit bodies are made by solvent-welding the conduit to the hub of the enclosure; or where the conduit enters an ordinary knockout, a connection is made by solvent-welding a threaded adapter to the conduit and using a metal knockout to secure it to the enclosure.

A variety of PVC elbows, boxes, and L- and T-shaped conduit bodies and service heads are available for use with PVC conduit. PVC must be heated for bending, which means the do-it-yourselfer is limited to using readily available factory items for turning corners and changing elevations. The 360-degree limit between openings also applies to PVC conduit. Supports must be no farther than 3 feet apart for ½- to 1-inch sizes, 5 feet apart for 1¼- to 2-inch sizes.

Rigid nonmetallic conduit is manufactured as Schedule 40 and as a heavier-walled Schedule 80 for use where greater physical strength is needed. PVC for use above ground is marked "sunlight resistant." Local practices determine which type is used where. Check with your electrical-inspection department or with a reliable local supplier before installation.

Flexible Conduit

Three types of flexible conduit are found in home wiring. They are flexible metal conduit, often called Greenfield, liquid-tight flexible metal conduit, and liquidtight flexible nonmetallic conduit. Special fittings are required for each of these types.

Greenfield is a fine, but expensive, general home-wiring material, once used exclusively in some geographical areas. It is pulled from box to box in concealed work through the holes bored in the framing of the house in much the same way cable is pulled. There can be no more than 360 degrees total in the bends between boxes, and the conduit must be supported at intervals not exceeding 4½ feet and within 12 inches of each box or panel. Properly installed Greenfield provides grounding at all the receptacles in the system.

Liquidtight flexible conduits in the home are usually limited to short lengths, such as between a disconnect and an air-conditioning compressor, or in wet or damp conditions where the lengths of conduit are no more than 6 feet and flexibility is needed.

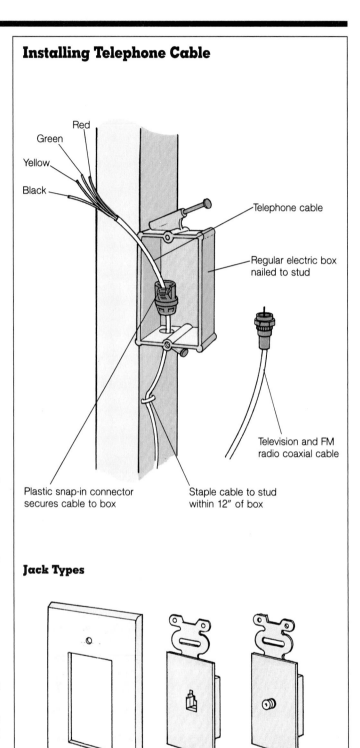

Installing Telephone Cable

Red
Green
Yellow
Black

Telephone cable

Regular electric box nailed to stud

Television and FM radio coaxial cable

Plastic snap-in connector secures cable to box

Staple cable to stud within 12" of box

Jack Types

Cover plate

Telephone jack

Cable jack

ROUGH WIRING IN REMODELING WORK

Remodeling presents a number of challenges not found in new construction. Work in newer one-story houses built above an unfinished basement isn't difficult because you can access all the walls by drilling up through the basement. Work in older homes is another story because you don't know what you will run into inside walls and ceilings.

Calculating Load

It's possible that your service is loaded to near capacity. The best way to determine this is to measure the current coming into the panel during a time of peak use—when dinner is baking in the oven, clothes are tumbling in the dryer, and the air conditioner is running, for example. This measurement is taken with a clamp-on ammeter, which simply clips over a conductor while the reading is being taken. However, taking this measurement is difficult, sometimes impossible, because there usually is little slack in the wires connected to the mains, and space inside the panel is tight. Also, few homeowners own or have access to an ammeter, so this job should be done by a professional electrician. Both hots should be read; the highest reading is the one to consider. Add the estimated load of the new work to the reading and compare that to the service rating (the capacity of the service entrance). Most likely there will be plenty of capacity available for the planned addition and, ideally, there will be space remaining in the panel for additional circuit breakers or fuses. However, if the readings plus the estimated load of the addition

approach the service rating, you may want to enlarge your service or downsize the planned addition. An alternate, but less accurate, way to estimate current is to add the total wattage of all light bulbs, permanent appliances, and anticipated loads during peak use. Divide this total by the voltage to obtain amperage. A rule of thumb to follow is that any circuit, the mains included, should not be loaded to more than 80 percent of capacity if the load is to continue for more than three hours at a time.

Tying in to Existing Circuits

The real difficulty in connecting new receptacles to existing wiring often lies in getting the cable to the panel or into existing outlet boxes where the connection will be made to the wiring system. Remember that all such connections must be made in boxes or at the panel. Once the cable is in the box or panel, the connections should not be difficult to make.

At the Receptacle

Before beginning any actual work, shut off the power to the outlet box from which you plan to feed the new receptacle.

Remove the cover and then remove the screws and pull the receptacle and its wiring as far out of the box as possible. Using a flashlight, take a long, careful look inside the box. If it appears to have just enough room for its receptacle and existing wires, you probably should go elsewhere for power. Also, you may find that you can't physically get another cable into the box because there isn't a knockout available where you need one.

Look at the top and bottom only. Don't consider coming into the back or sides of the box. If there is a knockout where you need one, test to see if it can be removed. Knockouts are intended to be removed from the outside of the box, so removing them from the inside is a difficult task at best. Sometimes no matter how hard you punch the knockout from the inside, it won't give. If this happens, try drilling a ¼-inch hole through the front edge of the knockout, prying it up with a screwdriver tip, and then twisting it out with pliers. This sometimes works. However, be aware that if it doesn't, you will have to close up the partial hole you have made. Any knockout that is removed accidentally has to be replaced with a closure.

The cable clamps in switch boxes are arranged to grip two cables coming into both the top and the bottom of the box. Some boxes have pryouts in the cable openings next to the clamps that can easily be removed from the inside with the tip of a screwdriver.

After looking at the conditions inside the box, you'll have to decide whether to tap the power here or go elsewhere. The ideal choice for powering new

receptacles is at the service panel, but getting cable back to the panel in an old home is often difficult because of the distances and obstacles involved.

Power can also be picked up from existing surface-mounted boxes in unfinished basements and attics, or from midspans of exposed cable runs using the two-box method described in the section on Knob-and-Tube Wiring later in this chapter. Don't forget to run a ground wire when this is done; it is very important.

At the Subpanel

Another solution is to install a subpanel. Subpanels are enclosures that contain a number of circuit breakers or fuseholders intended to feed small individual loads. They differ from the service panel in that they don't have a main breaker and the neutral bus bar is not bonded (electrically connected) to its enclosure. Instead, a separate bus is installed so that it is bonded to the enclosure as well as to the service panel. The bare grounding wires in all the cables leaving the subpanel connect to this bus.

Subpanels that supply power to additional branch circuits are often installed next to the service equipment when all the circuit-breaker spaces in the main panel have been used up. They're also used where several branch circuits are needed in a certain space such as a workshop or kitchen. A subpanel may also be installed in an area where extensive remodeling is to be done and several new branch circuits will be needed.

Mount the subpanel where it will be readily accessible

without having to move or climb over anything. You can surface-mount a subpanel in unfinished spaces, but it should be flush-mounted anywhere else in the house. Flush-mount the subpanel in inside walls for easiest access.

Adapting to Existing Wiring

Wiring to electrical outlets in remodel work is sometimes connected to nearby receptacles rather than being run all the way back to the panel. This is especially true if the remodeling involves adding only one or two receptacles, and the route to the panel is difficult. If possible, running a cable up from an unfinished basement or down from the attic is generally the easiest option.

Nonmetallic-sheathed cable is commonly used in remodeling even when another wiring method was used to wire the house originally. Connecting to the existing wiring usually means removing the existing receptacle connections, splicing them to the new cable, and adding a set of pigtails—6-inch pieces of wire that are used to reconnect the receptacle to the house wiring. Be sure to connect color to color, and attach the white pigtail to the silver-colored screw on the receptacle, the black to the brass-colored screw, and the bare or green to the green screw. Don't simply connect to the second set of screws on the receptacle; this would create a possible shock hazard.

Aluminum Wiring

Many of the homes built around 1970 were wired with aluminum wiring, either nonmetallic-sheathed cable or aluminum

wire in Greenfield, in order to save money. Aluminum wiring still costs about half what copper wiring costs, but it also has been the cause of many residential fires, according to the Consumer Product Safety Commission. Heating problems sometimes occur where the aluminum wire connects to a screw terminal on a switch or receptacle, often because the screw was overtightened during installation, seriously nicking or flattening the wire and reducing its cross-section area which in turn reduced its ampacity. Conductor oxidation at terminals can also create high-resistance connections that can cause heating problems.

One remedy for houses where aluminum wire was installed in Greenfield is to replace the aluminum with copper. This isn't as difficult as it may sound. With the power off, the aluminum wire is used to pull the copper wire in. Having a plan of the wiring scheme of the house makes this job a bit easier.

A simpler method for both aluminum nonmetallic-sheathed cable and aluminum wire in Greenfield is to turn off the power and disconnect the aluminum conductors from the switch or receptacle. Connect copper pigtails to the aluminum, and then reconnect the device to the copper wiring. Wire nuts used to be approved for use with copper-to-aluminum connections in dry locations, and this configuration is safe in existing wiring. However, the code specifies that wire nuts can no longer be used for copper-to-aluminum connections. There

are now special connectors available to do the job. They are designed in such a way that the copper and aluminum contact the connector material but not each other, so galvanic action, which corrodes metal, can't take place between them. A common connector for this application is a split-bolt connector with a separator that

keeps the two conductors apart. Another type is shaped like a cylinder, has a hole in each end for the wires, and uses setscrews to hold the conductors firmly in the connector. An antioxidant compound should be applied to all aluminum connections when they're being made in order to prevent

Aluminum and Knob-and-Tube Wiring

Knob and Tube

Knobs

Tubes

Aluminum

Switch or receptacle must be marked CO/ALR. Any other marking, such as AL-CU, is not acceptable.

Aluminum cable

Aluminum ground wire

Wires must be attached at screws, not push-in terminals

serious oxidation. These connectors must also be insulated with electrical tape before their circuits are energized again.

Knob and Tube

Rarely, a cable connection must be made to knob-and-tube wiring, a wiring method used when homes were first electrified. In this system, cylindrical porcelain knobs were nailed into position to support individual conductors along the sides and across the edges of framing members. When the wiring had to pass through a framing member, porcelain tubes were set in holes drilled at a slightly downward angle and the individual wires were pulled through the tubes. It's very difficult to connect new wiring to a receptacle in this type of system for a variety of reasons. It's likely that the wire insulation may be brittle with age; it may look intact, but touch it and it disintegrates. The best place to connect to a knob-and-tube system is in the attic along the floor joists, and this often means lifting some floorboards to uncover the wiring. Look first in the area above a receptacle. Wires feeding a receptacle, unlike a switch loop, are more likely to be on all the time.

You will also find that both wires in a knob-and-tube branch circuit are the same color, so your next task is to determine which is the hot wire and mark it with black tape. Start by turning off the power; then strip about 1 inch of insulation from one wire at a point midway between two knobs. To do this, simply crush the insulation with a pair of pliers, and then pick the bits of

insulation off the wire. Move downstream about 3 inches and do the same to the other wire. Next, turn the power back on and return to the wiring with a voltage tester. Hold a fingertip against one of the leads of the tester while you carefully hold the other test lead, without touching the metal tip, against the bare spot on one wire and then the other. Note that this procedure requires careful attention to safety. You must not be touching any other metal, standing in water or on damp ground, or grounded in any other way while performing this test. If you are uncomfortable with this technique, an alternative method is to ground one of the voltage-tester leads against a water pipe or other known ground (using a jumper wire if necessary) and touch the other lead to each of the wires you are testing. The hot wire is the conductor that lights your tester. Turn off the power again and mark the hot wire with black tape near the two closest knobs.

Now cut both wires midway between the knobs, mount a box within 12 inches of each of the knobs, and route the existing wires into their respective boxes. Connect the two boxes with a short piece of nonmetallic-sheathed cable and splice the wires to restore power to everything downstream. Power for additional outlets is now available in both boxes.

The porcelain bushings and flexible tubing (called loom) as well as the knobs, cleats, and tubes used in original knob-and-tube wiring have become antique-store items, so you'll have to innovate a bit. Use plastic 4-inch square boxes. Run the old wires into separate

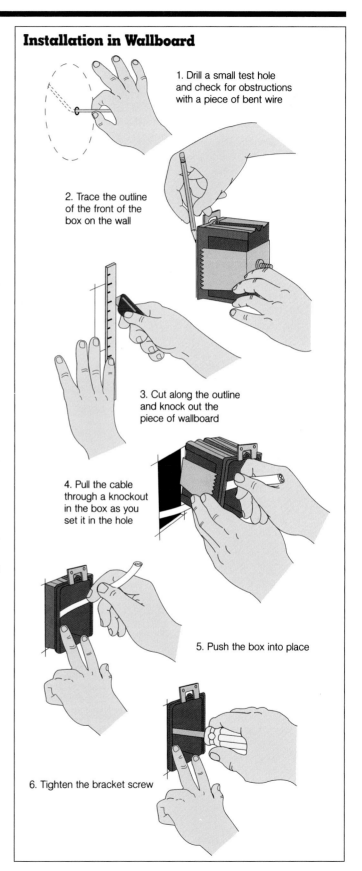

Installation in Wallboard

1. Drill a small test hole and check for obstructions with a piece of bent wire

2. Trace the outline of the front of the box on the wall

3. Cut along the outline and knock out the piece of wallboard

4. Pull the cable through a knockout in the box as you set it in the hole

5. Push the box into place

6. Tighten the bracket screw

Installing a Ceiling Box in Wallboard

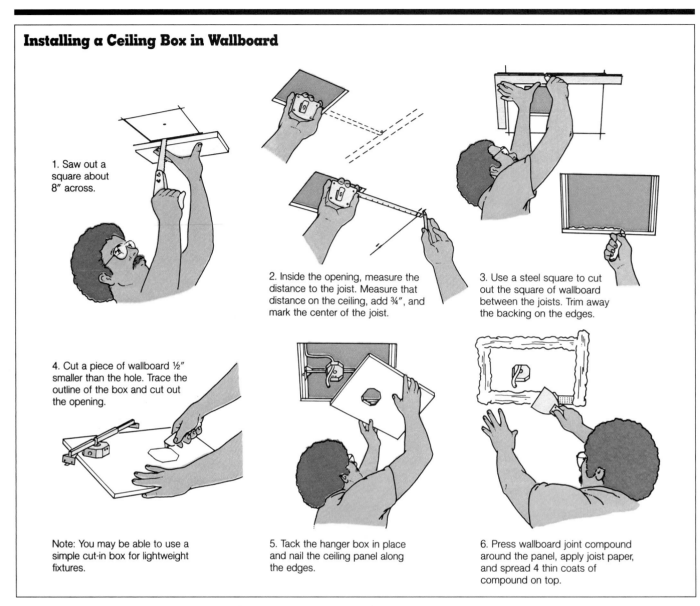

1. Saw out a square about 8" across.

2. Inside the opening, measure the distance to the joist. Measure that distance on the ceiling, add ¾", and mark the center of the joist.

3. Use a steel square to cut out the square of wallboard between the joists. Trim away the backing on the edges.

4. Cut a piece of wallboard ½" smaller than the hole. Trace the outline of the box and cut out the opening.

Note: You may be able to use a simple cut-in box for lightweight fixtures.

5. Tack the hanger box in place and nail the ceiling panel along the edges.

6. Press wallboard joint compound around the panel, apply joist paper, and spread 4 thin coats of compound on top.

openings in each box after removing the cable clamp from the end of the box. Secure the nonmetallic-sheathed cable with the clamps in the other ends of the boxes and strap the cable within a foot of each box. Remember, these boxes must remain accessible, so install them so the cover plates remain visible when you replace the floorboards.

Grounding

The method described above doesn't provide grounding for receptacles that will be added to this circuit. A ground fault circuit interrupter (GFCI) should be installed at the first electrical outlet fed from this tap in the attic, and any additional receptacles should be fed from the GFCI. Do not connect a grounding conductor to these additional receptacles. This way, all the new receptacles

will be deenergized if a ground fault occurs on any of them.

Planning Runs

Plan and do your work in ways that yield the desired results, yet strive for a minimum of cleanup and restoration work. Avoid work in outside walls except where absolutely necessary. It is often nearly impossible to push a fish tape up an

outside wall cavity because it is filled with batt-type insulation or because of the fire-stops and other obstructions often found in older homes. Route cables through inside walls whenever possible.

Drilling up from the basement or down from the attic is almost always easier than running cable through a finished wall. It's sometimes advantageous to run cable up through a closet. When this is done, the holes should be drilled in

Metal Lath and Plaster Walls

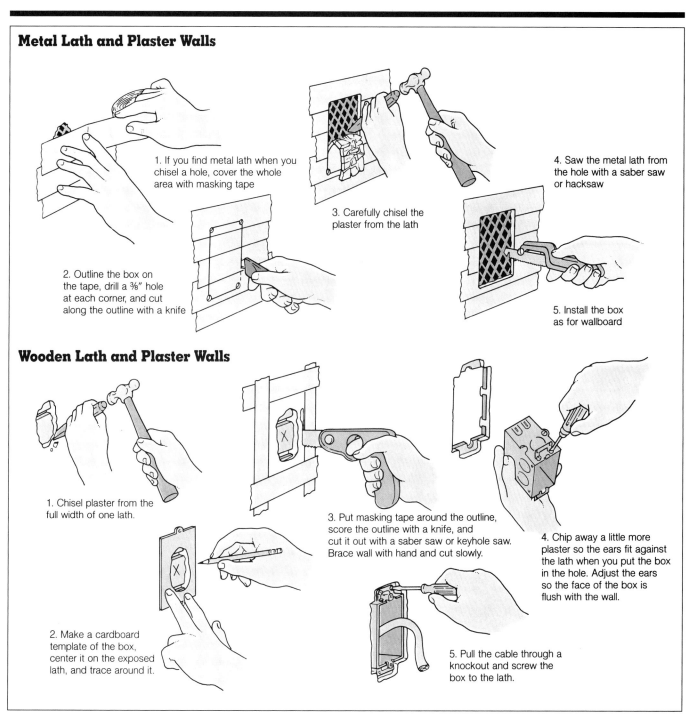

1. If you find metal lath when you chisel a hole, cover the whole area with masking tape

2. Outline the box on the tape, drill a ⅜" hole at each corner, and cut along the outline with a knife

3. Carefully chisel the plaster from the lath

4. Saw the metal lath from the hole with a saber saw or hacksaw

5. Install the box as for wallboard

Wooden Lath and Plaster Walls

1. Chisel plaster from the full width of one lath.

2. Make a cardboard template of the box, center it on the exposed lath, and trace around it.

3. Put masking tape around the outline, score the outline with a knife, and cut it out with a saber saw or keyhole saw. Brace wall with hand and cut slowly.

4. Chip away a little more plaster so the ears fit against the lath when you put the box in the hole. Adjust the ears so the face of the box is flush with the wall.

5. Pull the cable through a knockout and screw the box to the lath.

a corner of the closet and the cable must be protected by metal pipe or conduit where it is exposed. The easiest way to ensure protection is to use BX cable for the exposed portion. Nonmetallic cable can be fished in the space between the first and second floors if the route

runs parallel to the floor joists. Simply cut pockets in the floor next to the wall at both ends of the run and fish the cable from pocket to pocket. The pockets also allow you to drill down into the walls below. Restoring a floor is usually easier than

restoring the ceiling below, especially when the floor is carpeted.

Cut-in Boxes

There are a variety of boxes available for remodeling. The principle difference between

remodeling boxes and those used in new work is in the methods used to mount them: Utility boxes in new work are mounted directly to the house framing, whereas cut-in boxes in remodeling are attached to the finish material of the walls or ceiling, whether it is plaster, wallboard, or paneling. The plaster ears on metal cut-in boxes and the flanges on plastic cut-in boxes keep them from falling into the hole. Simply tightening a device or devices attached to the box clamps the box in place from the inside.

To mount a cut-in box, first cut a hole in the wall the approximate size and shape of the box. This cutout must be made at a point in the wall that is clear of any framing or other obstructions.

Usually a template, furnished with the box, is used to mark the shape of the opening. If you don't have a template, hold the box with its opening against the wall and trace the outline on the surface. Do not trace around the plaster ears or mounting flanges. After marking the desired shape on the wall, make the cutout using a method appropriate for the type of wall material involved.

Pull the cable up through the wall and into the opening until about 12 inches of cable sticks out of the wall. Then prepare the box to accept the cable. Remove the cable clamp and, if it is a metal box, the knockout that is going to be used. If the box is plastic, punch out the cable opening and trim it smooth with a knife. Loosely reattach the cable clamp and push the cable into the box through the knockout or cable opening, far enough so that at least 8 inches of the outer jacket of the cable can be removed. Then remove the jacket with pliers and a cable ripper or knife.

Take the cable end in one hand and the box in the other and, tilting the upper end of the box away from the wall, push the bottom of the box into the opening until the cable entering it is well inside the wall. Then straighten the box and push it into the wall as far as it will go. The plaster ears or mounting flanges will keep it from falling into the wall cavity. Push the excess cable down through the box and into the wall until just ¼ to ½ inch of the cable jacket shows above the clamp.

Tighten the cable-clamp screw, and then tighten the screw or screws that operate the mounting device on the box until the box is held firmly in the wall.

Snap brackets and swing clamps are two popular clamping devices found on non-metallic old-work boxes. A snap bracket is a U-shaped piece of sheet metal about 1½ inches wide that fits loosely over a box, covering both sides and the bottom of the box. A coarse-thread tapping screw passes through a hole in the back of the box and then into a tapped hole in the back of the bracket. The box is pushed all the way into the cutout in the wall, and then the bracket screw is pushed into the box, which pushes the snap bracket far enough into the opening for its ends to clear the opening and snap open about ½ inch on each side of the box. The bracket screw is then tightened until the box is held securely in place.

Swing clamps are flags, or fingers, that slide back and forth in channels molded on

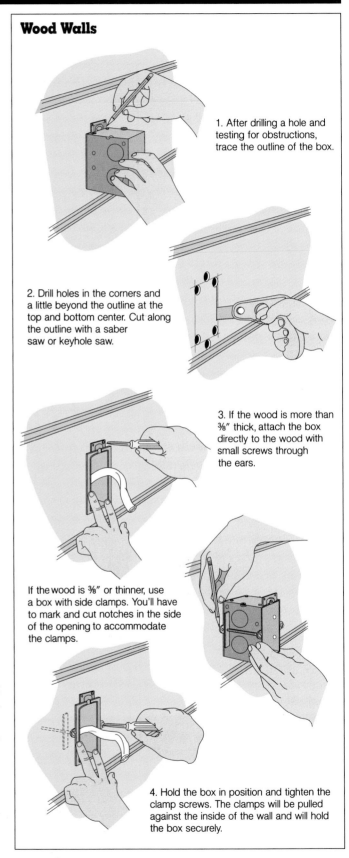

Wood Walls

1. After drilling a hole and testing for obstructions, trace the outline of the box.

2. Drill holes in the corners and a little beyond the outline at the top and bottom center. Cut along the outline with a saber saw or keyhole saw.

3. If the wood is more than ⅜" thick, attach the box directly to the wood with small screws through the ears.

If the wood is ⅜" or thinner, use a box with side clamps. You'll have to mark and cut notches in the side of the opening to accommodate the clamps.

4. Hold the box in position and tighten the clamp screws. The clamps will be pulled against the inside of the wall and will hold the box securely.

the outer sides of the box and are allowed to swing out 90 degrees from the sides of the box. Tapping screws pass through hubs on the front of the box, which are positioned directly over the channels, and then thread into the flags. The flags are against the box before it is pushed into the cutout. When the box is in place, tighten the clamp screws. The first quarter turn swings the clamps out 90 degrees from the sides of the box; the remaining turns draw the clamps toward the front of the box, thus clamping it firmly in place.

Sometimes cut-in metal boxes are installed using a box support that is nearly identical to the snap bracket described earlier. In addition, two other types of supports are commonly used with metal boxes.

One such type uses a jack-screw arrangement on each side of the box. Turning the screws causes the slotted, straight strips of metal which the screws thread into to form a V shape that pulls toward the front of the box. This clamping action will hold the box firmly in place.

Sheet-metal grip fasteners or box supports are used in pairs to mount metal boxes in wall materials up to 1 inch thick. (The other types described in this chapter are for use in walls that are approximately ½ inch thick.) The fasteners, sometimes called steamboats or mouse-traps, resemble the letter *F*, with a 1-inch extension of the vertical line of the letter. After the box is in place, a clamp, with the long leg up, is worked into the cutout next to the box. The clamp is centered vertically

on the box and pulled forward. While holding the box to keep it from shifting, bend each finger of the clamp sharply over the front edge and down into the box. Repeat the procedure on the other side of the box, then pinch all four fingers tightly with pliers.

Boxes can be installed in wood paneling or baseboards by attaching them to the wall with small wood screws that pass through holes in their plaster ears. Take care to make the cutouts no larger than absolutely necessary; if they are too large, there may not be enough wall material at the top and bottom of the box to accommodate the screws.

Metal boxes can be fastened to the wood lath in lath-and-plaster walls. The procedure begins with marking an approximate box location on the wall. Then, using a ½-inch cold chisel and hammer and a drill with a small masonry bit, chisel and drill a small, ½-inch-wide vertical channel through the plaster to the lath but not into it. Continue the channel, working a little bit up and then down, until one full lath is exposed. The center of this lath will be the center of the cutout. Draw an outline slightly larger than 3 inches high and 2 inches wide centered vertically over the center of the lath. Using the drill and chisel, remove all the plaster within this area and then carefully saw out the lath with a keyhole saw.

Test-fit the box and trim the opening where necessary. Slip the box into the cutout and trace the outline of the plaster ears on the wall. Take the box

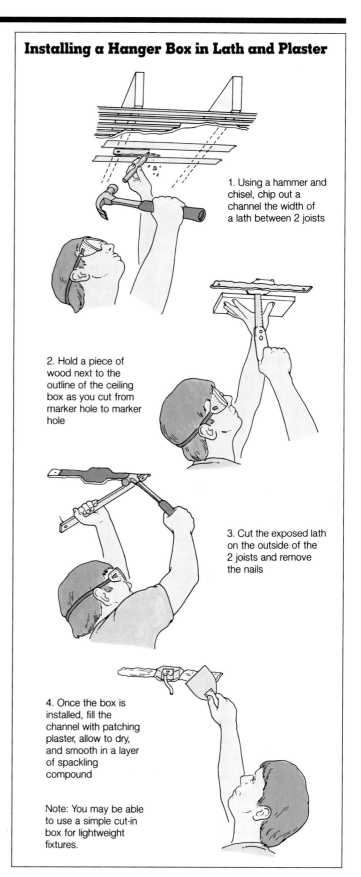

Installing a Hanger Box in Lath and Plaster

1. Using a hammer and chisel, chip out a channel the width of a lath between 2 joists

2. Hold a piece of wood next to the outline of the ceiling box as you cut from marker hole to marker hole

3. Cut the exposed lath on the outside of the 2 joists and remove the nails

4. Once the box is installed, fill the channel with patching plaster, allow to dry, and smooth in a layer of spackling compound

Note: You may be able to use a simple cut-in box for lightweight fixtures.

Running Cable in an Attic

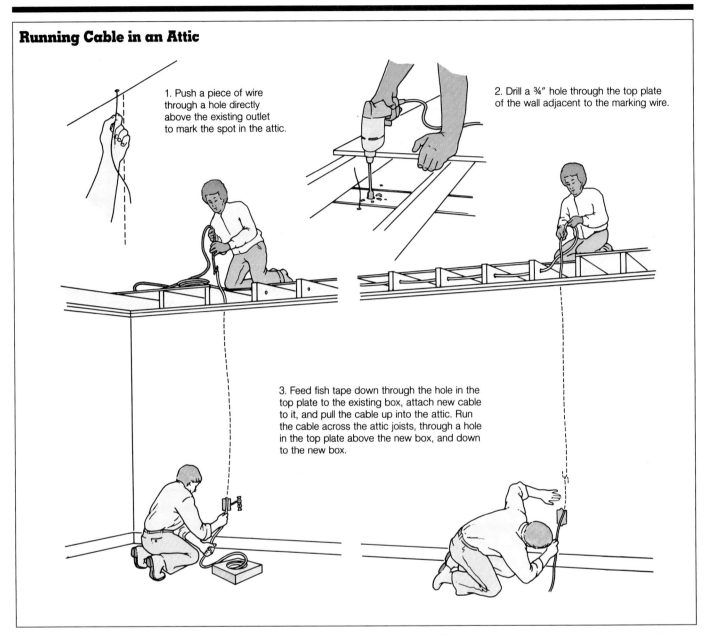

1. Push a piece of wire through a hole directly above the existing outlet to mark the spot in the attic.

2. Drill a ¾" hole through the top plate of the wall adjacent to the marking wire.

3. Feed fish tape down through the hole in the top plate to the existing box, attach new cable to it, and pull the cable up into the attic. Run the cable across the attic joists, through a hole in the top plate above the new box, and down to the new box.

out of the opening and remove only the plaster—not the lath—within the outlines. Adjust the plaster ears so they're set back from the front of the box the thickness of the plaster. Bring your cable into the box, and then work the box and cable back into the opening. Secure the box to the lath using small

screws run through the holes in the plaster ears. Because old lath splits easily, drill small pilot holes in the lath for these mounting screws. Fill the gaps around the box with spackling compound.

Another technique is to use the type of plastic box with a U-shaped spring clamp attached. You will have to cut off the ends of the clamp so it will clear the thick lath-and-plaster wall.

Running Cable Through Walls

It's usually easy to run cable on the surface or through unfinished walls. Difficulties arise when the cable leaves the basement or attic and enters the finished walls and ceiling spaces. Two special tools are required for working inside the walls: One is an electronic stud finder; the other is a fish tape.

The electronic stud finder lets you map the wall interiors without having to drill exploratory holes. You can locate studs, blocking, or other obstacles inside the wall without damaging it. Stud finders work very well on finish materials that are of uniform density, such as wallboard and paneling, and they work quite well on sound plaster walls.

One fish tape on the job is good; two is much better. One

Running Cable Under the House

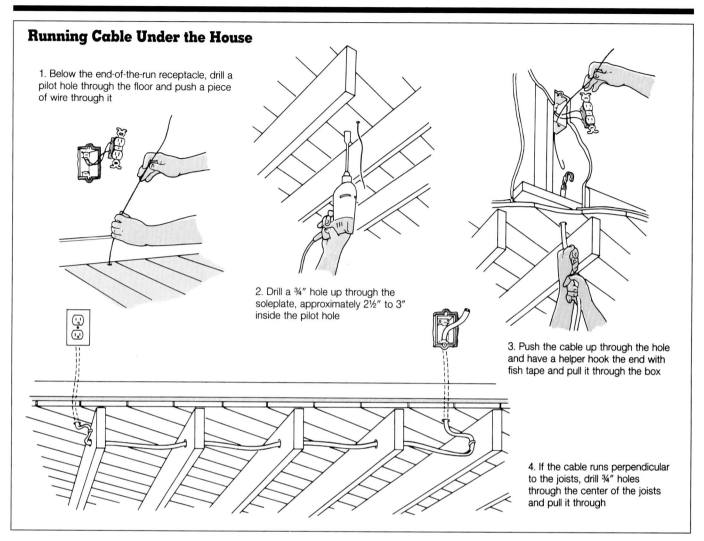

1. Below the end-of-the-run receptacle, drill a pilot hole through the floor and push a piece of wire through it

2. Drill a ¾" hole up through the soleplate, approximately 2½" to 3" inside the pilot hole

3. Push the cable up through the hole and have a helper hook the end with fish tape and pull it through the box

4. If the cable runs perpendicular to the joists, drill ¾" holes through the center of the joists and pull it through

fish tape lets you fish from a small or large opening into a larger area, for example, from a small hole through the top plates down into a cutout in the wall, or between pockets cut in the floor.

Most fish tapes used in residential work are made of flat metal spring stock that is ⅛ inch wide and ¹/₁₆ inch thick. Spring stock is used because it doesn't kink easily. It does, however, tend to arch, so were you to push it into a hole in the top plate with the intention of it continuing down inside the wall and into an open knockout in a box below, it would

never hit the knockout. A piece of solid wire with a small hook on the end can be pushed up into the wall through the open knockout to snag the fish tape and draw it out through the knockout, but success with this technique is marginal at best.

A more reliable method is to push the first fish tape down about halfway toward the box, and then push a second fish tape up through the knockout and most of the way up into the wall. Then the person working above cranks the top fish tape while the person

below moves the lower fish tape up and down. The two tapes will hook together and the upper tape can then be pulled down into the box.

Avoid damaging the ceiling if at all possible. Ceiling restoration is far more difficult than patching a wall or fixing most floor damage.

Closing up a hole in wallboard isn't difficult. Cut two pieces of 1 by 3 or 1 by 4 to a length about the height of the hole. Then, using a variable-speed drill, a Phillips-head bit, and wallboard screws, slip the wood strips into the hole and

attach one to each side by holding them in place and screwing through the wallboard. Half the width of each strip should be inside the wall and the other half should be exposed; put the cutout piece in place and screw it to the exposed half of each strip. Then tape, apply joint compound, and sand to finish the job. The cutout won't show.

Some ways of running cable through walls involve extensive cutouts and channeling. This technique adds a lot of difficult restoration work to the

Installing Ceiling Boxes from Above

1. Drill a ⅛" marker hole where you want the new box in the ceiling.

2. Using the marker hole and nails as a guide, cut a board or 2 from the attic floor.

3. Center a box on the marker hole, trace its outline, and drill holes at each of its 8 corners.

4. Cut from hole to hole to make an opening that exactly fits the new box.

5. Use metal shears to snip the tabs from the ends of the adjustable hanger. Adjust the hanger to fit between the joists, attach the box so it is positioned directly over the hole, and screw the hanger to the joists.

6. Nail 2×4 cleats to the joists so the floorboards will have something to rest on when you replace them.

job. Sometimes this is unavoidable of course, but it should be kept to a minimum. Alternative routes are often present behind baseboards and door casings.

Whatever the case, whenever you run across a wall rather than through it, remove the finish material down to the studs, and then put notches in the studs to lay the cable in. After the cable is in place, nail approved plates over each notch as protection against damage by nails or screws. Then close up the walls in appropriate fashion.

Some Typical Projects

Often a room needs an additional receptacle on a particular wall. If the location on the wall isn't critical, and if there is a receptacle somewhere on the other side of the wall in the next room, installing the new receptacle is no problem.

Here, as in most remodel work, you have to have an accurate sense of the framing inside the wall. Use a stud finder or other means to locate the studs in the area in which you will be working. Plot their locations with light pencil marks on the wall. With this information, you should be able to tell on which side of the stud the existing outlet box is mounted, and from this location determine how much space is available for another box at the same elevation in the same wall void.

Make a reference mark on the wall about 3 or 4 inches from the side of the void away from the existing box. Measure carefully along the wall from this point to a point that is common to both sides of the wall—a doorway, perhaps—and measure that same distance back on the other side of the wall. This point will be the centerline of the cutout for the new box.

From here it is just a matter of making the cutout, installing the cable, and connecting it at both ends. Remember, when picking up power from an existing receptacle, the original wires should be disconnected and spliced to the new cable along with a set of pigtails that will be used to reconnect the old receptacle to the house electrical system. Don't use one receptacle to directly refeed another; this could create a possible shock hazard.

Another frequent remodel project is installing a new ceiling light and switch in a room. If the ceiling isn't accessible from above, determine the direction the ceiling joists run because the wiring to the ceiling box must be fished in the same direction. Then mark a location for the box that falls between two joists, and cut an opening in the ceiling sized for the cut-in box you will be using. For boxes with hanger bars you will need to cut the opening large enough to expose both ceiling joists. Now go to a point on an inside wall beneath the two joists the light will be between and, beginning about 4 inches down from the ceiling, make a clean rectangular opening in the wall. It should be about 16 inches high and 12 inches wide. Once inside the wall, drill a ¾-inch hole straight up through the top plates into the ceiling space. Form a smooth bend in the end of a fish tape and push it up into the hole so that the hook is heading more or less toward the opening in the ceiling. Push until you think there is enough tape in the ceiling to reach the cutout. Then go to the cutout, reach inside, and pull out the tape. Pull cable from the ceiling box through the opening in the wall, and then route it to the new switch location. If you are not bringing power to the light fixture through the switch, you will have to fish a new cable from the power source to the ceiling fixture in a similar fashion.

When working in walls covered with wallpaper, cut three-sided flaps in the paper and tape them up out of the way wherever you need to cut a hole in the wall. Make the flaps large enough to fully cover your repair work in the wall.

Surface Raceways

Sometimes it is not practical to extend a circuit using any of the old-work methods described earlier. When this is true, surface raceways, metal or plastic, offer a low-cost, easy-to-install option. These raceways, which have been available for years, avoid the messy, time-consuming, and somewhat difficult job of opening walls and

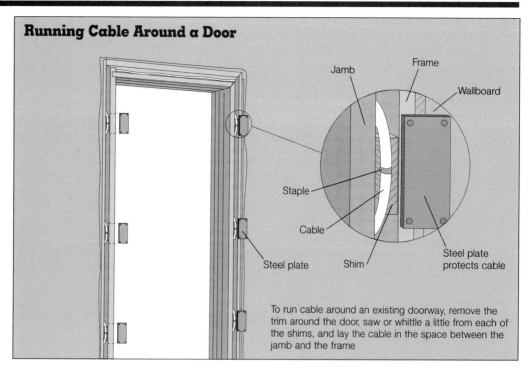

Running Cable Around a Door

Jamb · Frame · Wallboard · Staple · Cable · Steel plate · Shim · Steel plate protects cable

To run cable around an existing doorway, remove the trim around the door, saw or whittle a little from each of the shims, and lay the cable in the space between the jamb and the frame

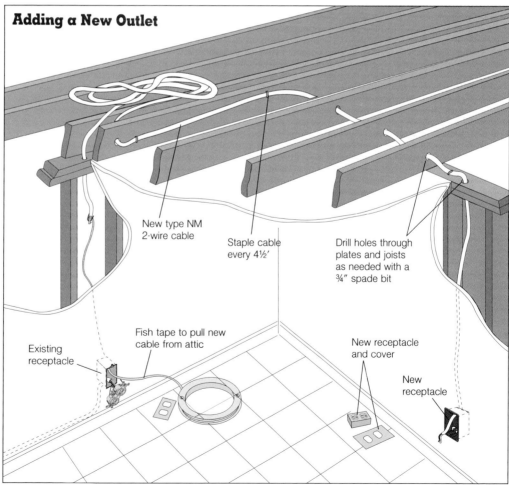

Adding a New Outlet

New type NM 2-wire cable

Staple cable every 4½'

Drill holes through plates and joists as needed with a ¾" spade bit

Existing receptacle

Fish tape to pull new cable from attic

New receptacle and cover

New receptacle

Raceways and Plug-in Strips

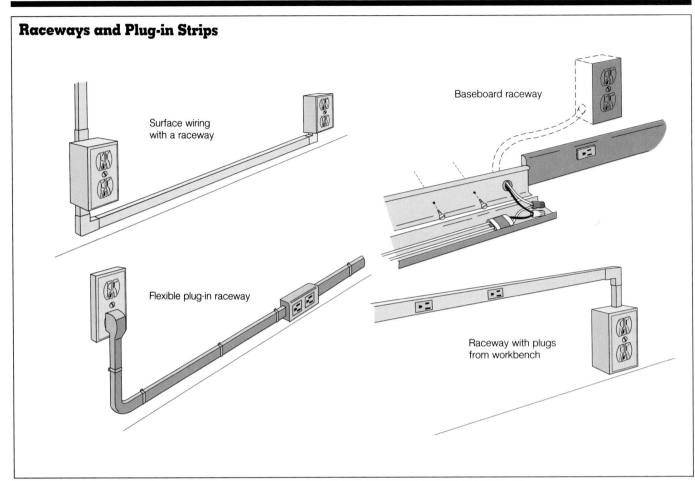

Surface wiring with a raceway

Baseboard raceway

Flexible plug-in raceway

Raceway with plugs from workbench

Running Cable Behind Walls

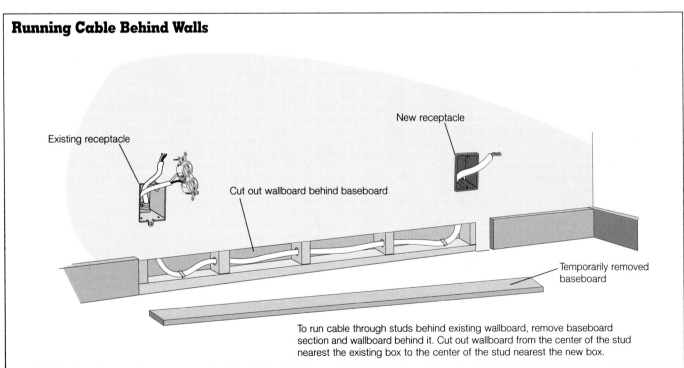

Existing receptacle

New receptacle

Cut out wallboard behind baseboard

Temporarily removed baseboard

To run cable through studs behind existing wallboard, remove baseboard section and wallboard behind it. Cut out wallboard from the center of the stud nearest the existing box to the center of the stud nearest the new box.

Switch and Ceiling Light

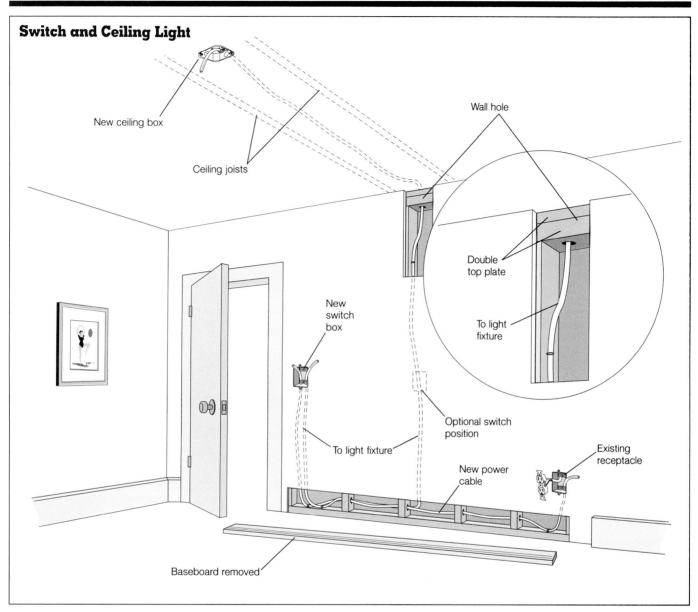

New ceiling box

Ceiling joists

Wall hole

Double top plate

To light fixture

New switch box

Optional switch position

To light fixture

New power cable

Existing receptacle

Baseboard removed

ceilings. Their main drawback is that they sit exposed on wall and ceiling surfaces. The system includes outlet starter boxes that mount over existing receptacles and are used to provide power for the system, and a variety of switch boxes and lighting outlet boxes.

The boxes are mounted on the surface and are connected by channel through which the raceway-system wiring is routed. Elbow fittings are used to make turns where needed.

Another popular system also uses an outlet starter box but doesn't use switch boxes. Rather, a wider channel is connected to the starter box, and special receptacles mount directly in the channel. The areas between receptacles are protected by a flat cover that simply snaps into the channel. This system is especially appropriate where many receptacles are needed in a small space, such as along the back of a workbench.

Permits and Inspections

Remodel work requires permits and inspections in much the same way new work does. Discuss your project when applying for your permit, and an inspector will probably adapt the inspection schedule to the complexity of your job. If the project involves just a receptacle or two fished up from the basement, the inspector will probably only want to look at it once before the receptacle is energized. However, if the cable runs require complex channeling or extraordinary routing, an inspection may be necessary before any restoration work begins. Then there will be a final inspection after the restoration work is done and the wiring is completed.

OUTDOOR ROUGH WIRING

Taking electrical wiring outdoors allows the installation of lighting, receptacles for powered equipment, and pool and spa systems. Outdoor materials must be protected from weather, foot traffic, and digging.

Setting Up Boxes

The weatherproof metal boxes used outdoors are made of a lightweight metal alloy rather than the thermoplastic or metal used in interior switch boxes. Outdoor boxes have wiring entry holes tapped for a common conduit size, usually ½ inch, in both ends and the back. Conduit threads to the box, as do cable connectors. Unused openings are closed with plugs furnished with each box. Boxes are sometimes supported by the rigid metal conduits threaded into them, or they're mounted with screws on a firm surface.

Ordinary switches and receptacles are used outdoors. They are protected from weather by the integrity of the box in which they are mounted and by the design of the gasketed covers used with them. Outdoor-receptacle covers commonly have a pair of spring-loaded gasketed doors that close over receptacle openings when not in use. GFCI covers have a single door. The switch cover usually has a small shaft passing through it. A fork-shaped device that slips over the switch handle is connected to the shaft on the inside of the cover, and an operating lever is attached to the shaft on the outside. This design prevents water from entering the box through the cover opening.

Plastic outdoor conduit systems use PVC boxes and gasketed covers.

Generally, boxes should be mounted on rigid surfaces rather than depending only on conduit for support. This protects both the boxes and the conduit from damage. It's also much easier to mount the boxes this way. The *NEC* allows conduit-only support of boxes that contain receptacles or switches only when two or more conduits are threaded into the box and the box is no more than 18 inches above the ground. Consider using pressure-treated 4 by 4s to support your boxes. They can be set to any height.

Underground feeder (UF) cable should be the first choice for in-the-ground wiring. It's approved for direct burial in the earth and in most cases requires mechanical protection only where it enters or leaves the ground. Both PVC and rigid metal conduit provide that protection. A bushing is required on the end of the conduit that is in the ground. If for some reason you choose, or are required to use, conduit in the ground, consider PVC first. It's fast and easy to install, and installation doesn't require special tools. PVC conduit and fittings also cost far less than rigid-metal-conduit systems.

Outdoor Electrical Plan

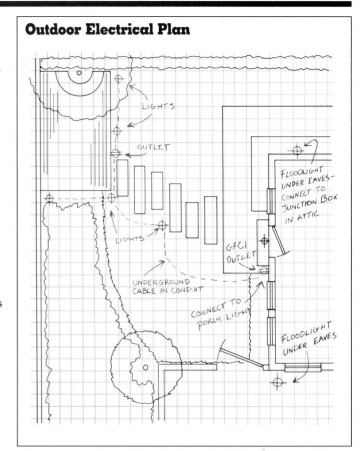

Burial and Overhead Rules

A number of *NEC* rules govern the installation of underground wiring. Cables for 15-amp and 20-amp 120-volt branch circuits can be buried in a 12-inch-deep trench if the circuit has GFCI protection. Landscape lighting that operates at no more than 30 volts requires a trench just 6 inches deep when wired with Type UF or other identified cable. Rigid metal conduit also requires a 6-inch-deep trench.

Deeper trenches are required for larger branch circuits and feeders unless rigid metal conduit is used. A 60-amp feeder to a garage, for example, would require a 24-inch burial depth for direct-burial cables or conductors, and

an 18-inch-deep trench if PVC conduit was being used. Where trenching is impossible—in solid rock, for example—rigid metal conduit or PVC conduit can be run across the rock surface if covered with 2 inches of concrete. Remember that wherever cables emerge from the ground, they must be protected from the bottom of the trench to a point 8 feet above grade, and that a bushing must be on the end of the protective conduit that is in the ground. Use rigid or Schedule 80 PVC for this protection.

Always leave some slack in the cable that is being put in the trench, and leave a small *S* in the cable at the point where the cable enters the protective conduit.

Extending Cable or Conduit Underground

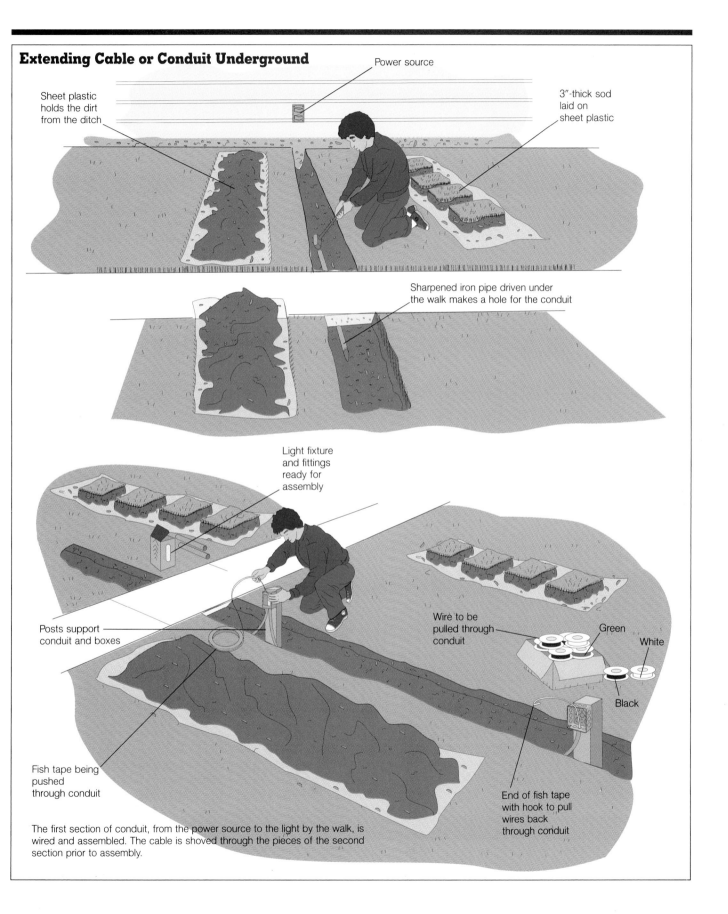

Power source

Sheet plastic holds the dirt from the ditch

3"-thick sod laid on sheet plastic

Sharpened iron pipe driven under the walk makes a hole for the conduit

Light fixture and fittings ready for assembly

Posts support conduit and boxes

Wire to be pulled through conduit

Green

White

Black

Fish tape being pushed through conduit

End of fish tape with hook to pull wires back through conduit

The first section of conduit, from the power source to the light by the walk, is wired and assembled. The cable is shoved through the pieces of the second section prior to assembly.

Outdoor Fixtures

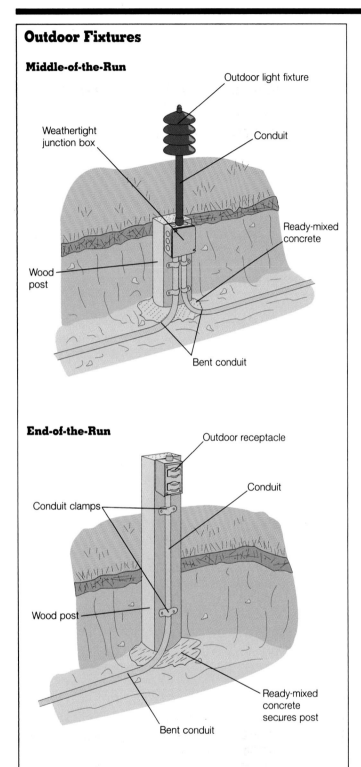

Middle-of-the-Run

Outdoor light fixture

Weathertight
junction box

Conduit

Ready-mixed
concrete

Wood
post

Bent conduit

End-of-the-Run

Outdoor receptacle

Conduit clamps

Conduit

Wood post

Ready-mixed
concrete
secures post

Bent conduit

Outdoor Wiring Fixtures

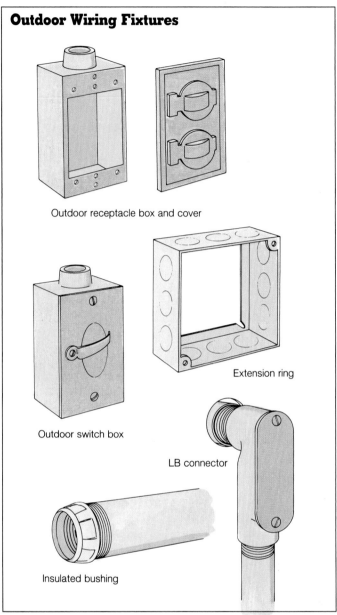

Outdoor receptacle box and cover

Extension ring

Outdoor switch box

LB connector

Insulated bushing

Trenches are usually backfilled with the material that was excavated. If that material is ordinary dirt or sand, there is no risk in simply laying cable in the trench and backfilling without providing extraordinary protection for the cable. However, protection is required where the presence of sharp rocks in the trench bottom and the backfill material may damage the cable. A bed and cover of sand, or running boards, will provide adequate protection. Also, if the trench passes through a garden or other area where future digging may occur, it is smart to cover the cable with boards before backfilling. The inspector will want to see the trench before it is backfilled.

For overhead services or feeders run overhead from the house to a detached building, such as a garage or workshop, service conductors must be run 10 feet above the yard and 15 feet above residential driveways. Branch circuits and feeders run overhead must be 10 feet above the ground and 12 feet above the driveway. Wires passing over roofs that can be walked on easily must pass at least 8 feet above the roof; other roofs require just 3 feet of clearance. Open conductors should not pass closer than 3 feet, measured horizontally, to windows and doors. If ever you add an open porch or deck to the house, these required clearances should be maintained.

Trenching Tricks

Trenching is usually done with a shovel, but when the trench is long or the ground is hard, think about renting a trencher. Route the trench well around large trees and other obstacles. More often than not, a trench must pass under a sidewalk or driveway at some point. Even when it is possible to dig under the pavement, it is usually unwise to do so. A good alternative is to drive a pipe through the ground at this point.

The pipe should not be too big; usually ¾-inch rigid conduit is large enough for No. 14 or No. 12 two-conductor cable. It should be about one-and-a-half times the width of the pavement you are passing under. Drive the pipe with a 3-pound engineer's hammer. The short handle and good balance

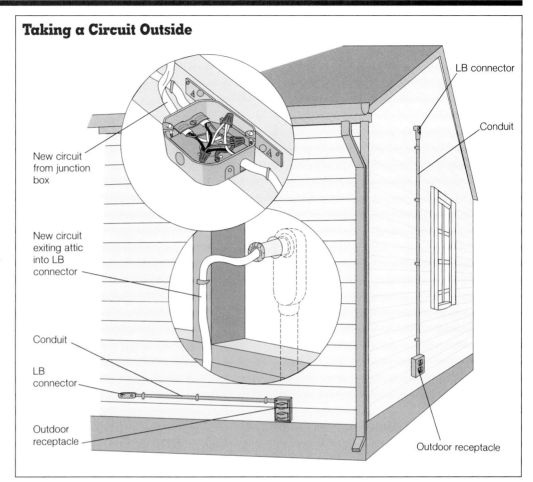

Taking a Circuit Outside

LB connector

Conduit

New circuit from junction box

New circuit exiting attic into LB connector

Conduit

LB connector

Outdoor receptacle

Outdoor receptacle

of this tool make it easy to use when you are kneeling next to the trench. It's also advisable to have a 14-inch pipe wrench on hand for rotating and extracting the pipe from the hole.

If you run into problems because of the width of the walk or difficulty with the soil, drive the pipe until it bogs down. Then, using a water hose and pistol-grip nozzle, make the tightest possible connection between the nozzle and pipe, and fully turn on the water briefly. Drive some more, then alternate between driving and water blasting until the pipe goes through. Don't use any more water than necessary; too much could cause serious undermining.

Wiring Pools and Spas

Without adequate safeguards, water and electricity can be a lethal combination. That is why the *NEC* rules for wiring swimming pools, the equipment used with them, and the areas immediately adjacent to them are numerous and complex. To detail the rules here would be beyond the scope of this book. Here is a brief review of the most important points, which may be enough to convince you that you probably should hire a licensed electrician to do the job. Your pool

installer can recommend an electrician who is well versed in wiring swimming pools.

All metal parts within or attached to the pool and within the area 5 feet from the pool walls, measured horizontally for a distance of 12 feet above the pool, must be bonded to a common bonding grid. Metal piping, pump motors, metal-sheathed cable and raceways, and all electrical equipment associated with the pool water-circulation system must also connect to this grid.

In addition to this bonding, the electrical equipment in and near the pool, and any pool-associated motors, must be connected to an equipment

Low-Voltage Outdoor Lights

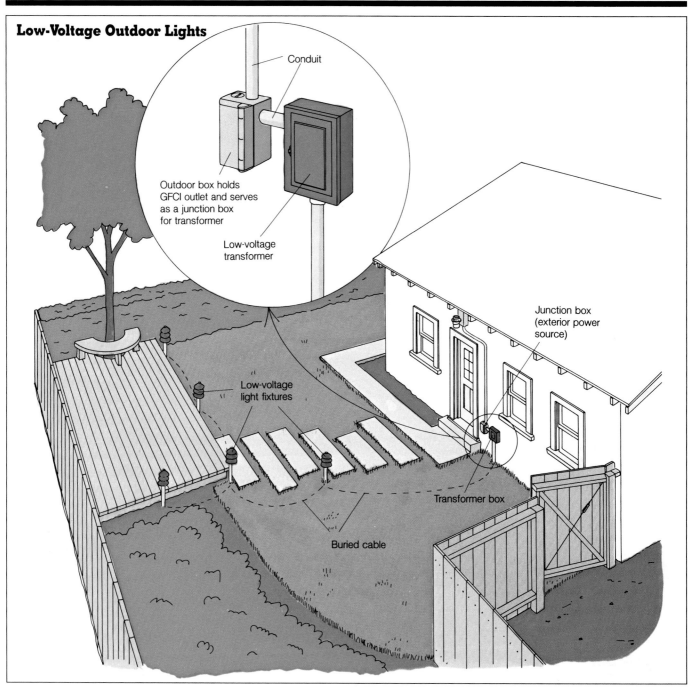

Conduit

Outdoor box holds
GFCI outlet and serves
as a junction box
for transformer

Low-voltage
transformer

Junction box
(exterior power
source)

Low-voltage
light fixtures

Transformer box

Buried cable

grounding conductor. Pump motors that are clearly marked "double insulation" are exempt from this regulation.

Ground fault circuit interrupter (GFCI) protection must be provided for all branch circuits used in or near the pool.

Lights and switches cannot be installed within 5 feet of the pool, and receptacles must be at least 10 feet away. However, one receptacle must be installed within 20 feet of the pool. There can be no underground wiring within 5 feet of the edge of the pool, and a pool cannot

be installed beneath an overhead service.

If you want to do the job yourself, professional pool installers should be able to help you with local regulations and practices. Make copies of Parts

A and B of Article 680 from a library copy of the Code to study at home. Work with your local inspector; if the job isn't going well, the worst thing that can happen is the inspector will recommend you hire a contractor to finish the job.

Minimum Cover* Requirements, 0 to 600 Volts, Nominal, Burial in Inches

Location of Wiring Method or Circuit	Type of Wiring Method or Circuit				
	Direct Burial Cables or Conductors	Rigid Metal Conduit or Intermediate Metal Conduit	Rigid Nonmetallic Conduit Approved for Direct-Burial Without Concrete Encasement or Other Approved Raceways	Residential Branch Circuits Rated 120 Volts or Less With GFCI Protection and Maximum Overcurrent Protection of 20 Amperes	Circuits for Control of Irrigation and Landscape Lighting Limited to Not More Than 30 Volts and Installed With Type UF or in Other Identified Cable or Raceway
All locations not specified below	24	6	18	12	6
In trench below 2"-thick concrete or equivalent	18	6	12	6	6
Under a building	0 (In raceway only)	0	0	0 (In raceway only)	0 (In raceway only)
Under minimum of 4"-thick concrete exterior slab with no vehicular traffic and the slab extending not less than 6" beyond the underground installation	18	4	4	6 (Direct burial) / 4 (In raceway)	6 (Direct burial) / 4 (In raceway)
Under streets, highways, roads, alleys, driveways, and parking lots	24	24	24	24	24
One- and two-family dwelling driveways and parking areas, and used for no other purpose	18	18	18	12	18
In solid rock where covered by minimum of 2" concrete extending down to rock	2 (In raceway only)	2	2	2 (In raceway only)	2 (In raceway only)

* Cover is defined as the shortest distance measured between a point on the top surface of any direct-buried conductor, cable, conduit, or other raceway and the top surface of finished grade, concrete, or similar cover.

Note 1. For SI Units: one inch = 25.4 millimeters
Note 2. Raceways approved for burial only where concrete encased shall require concrete envelope not less than 2" thick.
Note 3. Lesser depths shall be permitted where cables and conductors rise for terminations or splices or where access is otherwise required.
Note 4. Where one of the conduit types listed in columns 1–3 is combined with one of the circuit types in columns 4 and 5, the shallower depth of burial shall be permitted.

Reprinted with permission from NFPA 70-1990, the *National Electrical Code*, copyright © 1989, National Fire Protection Association, Quincy, MA 02269. This reprinted material is not the complete and official position of the National Fire Protection Association on the referenced subject, which is represented only by the standard in its entirety.

Low-Voltage Outdoor Lighting Systems

For inexpensive, easy-to-install lighting in a variety of forms for gardens, along walks, on decks, and so forth, low-voltage outdoor lighting kits are a sensible choice. The heart of the system is the power supply, which plugs into a 120-volt receptacle and supplies a safe low voltage, usually 12 volts, to power the lights. Some power supplies feature an electric eye that turns the lights on at dusk and off again at dawn. Installation is easy because the cable used in the system doesn't have to be run in a trench, although it can be, and the fixtures are mounted on stakes that simply push into the ground.

Connections at the fixtures are made by pins that pierce the cable insulation when the cable is positioned between the fixture head and stake and the two parts are forced into their final location and pinned there. When assembling the system from components, be certain the intended power supply is large enough. Add up the lamp wattages, then choose a power supply that offers at least 25 percent more wattage.

Don't use extension cords for permanent installations. It's much better to install 120-volt receptacles in areas where the lighting system is planned to be established.

THE MAST AND MAIN SERVICE PANEL

As a home electrician you will probably never have occasion to install a complete service panel yourself. However, at some time you may need to wire a new branch circuit into an existing panel. To accomplish this successfully you need to know how to evaluate the adequacy of the panel and, if necessary, how to locate and install a subpanel or how to replace an existing but outmoded panel even when the rest of the wiring for the home is all right. You also need to know how to calculate the shortest, most efficient branch–circuit runs from the panel. For any of these tasks it pays to know how service panels are installed, at least in theory, and for general electrical safety it pays to know how they operate in relation to the entire home electrical system. This chapter will guide you through this phase of the work, emphasizing the techniques, materials, and tools needed to do the job.

Installing the wires in the service panel in a neat and orderly fashion helps to keep all the circuits organized.

THE MAIN CONDUCTORS

In wiring an entire house, the next step after rough wiring is installing the main service panel. This ubiquitous metal box is the place where the electricity coming into the house connects with the branch circuits. It is the common denominator in any wiring job, whether it is adding a single new circuit to an existing service or wiring a whole house.

The Service Drop and Service Lateral

Power from the electric utility company's distribution system is brought to a house in one of two ways: either by overhead wires through what is called the service drop, or by underground cables through what is called the service lateral. The service drop connects to the service-entrance conductors outside the house at a point near the top of the service conduit. The other ends of the service-entrance conductors connect to the top terminals in the meter socket. The service-lateral conductors, although they enter the meter box from the bottom, nevertheless connect directly to the top terminals in the meter socket, except the neutral conductor, which runs unbroken to the neutral bus. From this point on, the two services are usually alike.

The terminals at the bottom of the meter socket connect to the service disconnecting means, and from there to the service overcurrent-protection device. In modern panels the disconnecting means and

the overcurrent-protection device are usually combined in the main service breaker. This breaker supplies power to the hot buses in the panel to which the branch-circuit breakers or fuses are connected.

Overhead Service

Service-drop conductors, including the required drip loops, must be at least 10 feet above grade at the building and in areas accessible only to pedestrians, and 15 feet above residential driveways. Also, service conductors must be installed out of reach of windows, doors, and porches, which according to the Code means beyond 3 feet. The area directly above a window is considered to be out of reach, so there is no restriction to wires above the top of a window.

The point at which the service drop attaches to the building should be below the top of the service conduit. The 10-foot-clearance rule is a major factor used to determine how the service conduit will be installed. It's often not possible to maintain this clearance on one-story houses without passing the service conduit through the roof. This type of installation is known as a low-roof service.

Low-roof service masts are always made of 2-inch rigid metal conduit because it has the strength necessary to support the service drop that attaches to it.

EMT (electrical metallic tubing) and rigid nonmetallic conduit, as well as rigid metal conduit, can be used when the service conduit doesn't have to pass through the roof to provide the necessary clearance and instead the service drop attaches directly to the house. In this case, the conduit can be smaller than 2 inches; 1¼-inch conduit is large enough for 100-amp services.

Check with your power company before installing the service equipment to make certain its service drop will attach to your house where you want it to. This usually isn't a problem unless the distance from the utility pole to the point on the house selected for the service is too great, or if for some reason the required vertical clearance can't be maintained.

Usually the service conduit rises straight up from the meter socket, but sometimes it must be bent 90 degrees to reach the point at which the service drop attaches to the house. In rare cases, because of the presence of porches, windows, and other obstacles, the route the conduit must take from the meter to the service drop requires a number of bends. Remember that the total bends between conduit openings can't exceed 360 degrees. If the installation requires more bends than permitted, use a pull box or C-conduit body in the middle of the run,

if necessary, to provide a midrun conduit opening and conform to the 360-degree requirement. Use the openings for pulling wire only; service-entrance conductors cannot be spliced in a pull box.

The exact location of the service conduit and meter socket is usually determined by the placement of the service panel. When the main circuit breaker is located in the service panel, the panel must be installed as near as possible to the point at which the service conductors enter the building. Service panels are commonly installed in basements or attached garages and, sometimes, on the outside of the house at a place where there will be sufficient working space around the panel and where it won't be necessary to move or climb over obstacles in order to reach it. If the panel is to be located well inside the building, the main breaker, or fused disconnect, must be located outside the house so that all the wiring inside will be properly protected from overcurrents and is accessible for emergency shutoff.

Read The Meter and Main Circuit Breaker in the next section to help you decide where the main breaker and perhaps the service-entrance panel itself will be installed. The meter box will need to be in place before conduit is installed to ensure that the two connect properly.

Installing the Mast

Once you have installed the meter box and main breaker, measure from the meter socket to a point a little more than 13 feet above grade. Cut a piece of 1¼-inch rigid metal conduit to

Service Drop

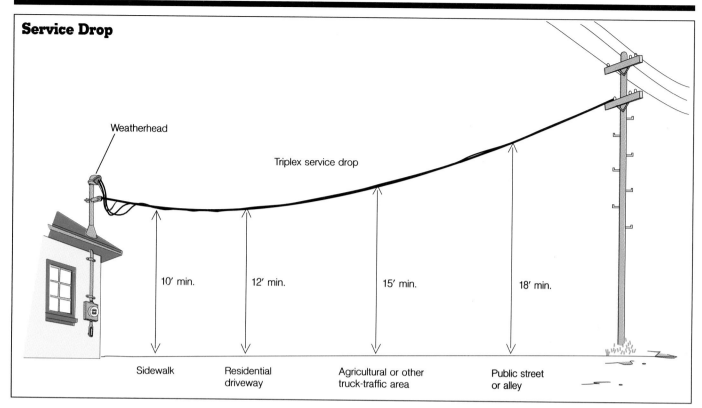

Weatherhead

Triplex service drop

10′ min.   12′ min.   15′ min.   18′ min.

Sidewalk   Residential driveway   Agricultural or other truck-traffic area   Public street or alley

this length. (For through-the-roof installations that require 2-inch rigid conduit, keep reading.) Ream or file the inside of the cut end of the conduit. Threading isn't necessary because the service-entrance head will simply slip over the conduit and will be held in place with setscrews or a clamp. Screw the weathertight hub onto the other end of the conduit.

Hold the conduit and hub centered over the top conduit opening of the meter enclosure. Plumb the conduit and make light pencil marks on the wall on each side of the conduit about 2 feet up from the socket and about 1 foot down from the top. Remove the conduit and fasten conduit hangers to the wall between the pencil marks, both top and bottom. Put the conduit back in place,

making certain the gasket is in place between the hub and the enclosure, and bolt the hub to the meter box. Secure the conduit in the hangers with screws and nuts. Slip the service head onto the conduit and tighten it in place. Mount a wire hanger or rack to the side and 1 foot below the service head. If triplex cable will be used for the service drop, install a single triplex wire hanger. If the drop will use three separate conductors, you will need three separate insulated wire hangers. Install the top wire hanger 1 foot from the top of the mast and the others approximately 6 inches apart.

A low-roof service is installed in much the same way except 2-inch rigid metal conduit is used and the holes through which the conduit will pass must be accurately cut in the roof and soffit. Special roof flashing and a conduit boot are

also necessary to do this job. Mark a plumb line that extends from the center of the socket hub to the soffit. Using a square, extend this line out to the edge of the soffit and make a pencil mark there. Then calculate the distance from the wall to the centerline of the conduit and mark this on the soffit with a line about 3 inches long that crosses the plumb-line extension. There is an easy way to determine how far out this will be. Simply measure from the back to the center of the conduit opening in the conduit hangers you are using.

Lining up the hole in the roof with the hole in the soffit can be done in several ways. An 18-inch-long ¼-inch drill bit can be used to drill a pilot hole up through the soffit and

roof. The drill must be held plumb side to side and front to back when this is done. Another method requires two people, very careful measuring, and a square and level. With this method, bring the plumb-line extension out to the edge of the roof by placing a square against the wall so that its other leg crosses the conduit centerline. Mark the edge of the fascia here. Measure the distance from the centerline to the fascia. Hold a 2-inch-long piece of 2 by 4 with square-cut ends flat against the fascia so that the bottom edge lines up perfectly with bottom of the fascia and one end is lined up with the centerline mark. This puts the open face of the 2 by 4 beyond the shingles. Add 1½ inches to the measure you just made and mark this dimension on the level (or on a short piece of wood if the level is not long enough). Next hold the square

Mast Installation

Service Drop

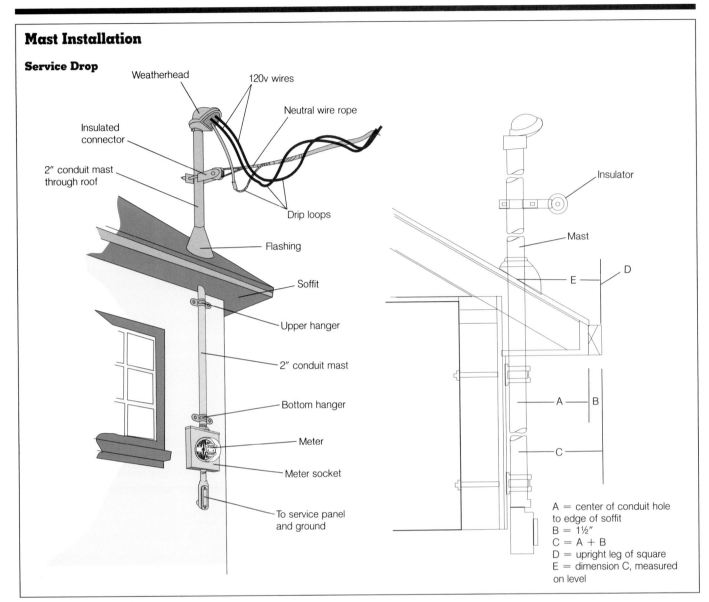

Weatherhead

120v wires

Neutral wire rope

Insulated connector

2" conduit mast through roof

Drip loops

Flashing

Soffit

Upper hanger

2" conduit mast

Bottom hanger

Meter

Meter socket

To service panel and ground

Insulator

Mast

D

E

A — B

C

A = center of conduit hole to edge of soffit
B = 1½"
C = A + B
D = upright leg of square
E = dimension C, measured on level

flat against the face of the 2 by 4 so that one leg is sticking straight up; eye it with the end of the 2 by 4 to make certain it is straight. Then hold the dimension mark on your level (or your board) against the upright leg of the square and level it where the far end touches the roof and the dimension mark touches the square. Hold the level or board as squarely as possible to the edge of the roof. Mark this spot by driving a nail

partway into the roof. Drill ¼-inch pilot holes at the marks, both up and down.

Once the pilot holes have been drilled, the conduit holes can be cut out with a 2½-inch hole saw or dial saw. Don't be too concerned if the two holes don't line up perfectly; the top hole can be opened up with a saw if necessary.

Install one conduit hanger about 1 foot above the meter socket and one 1 foot below the soffit. Then slip the flashing up under the shingle above the

hole (the bottom of the shingle may have to be cut out). Let the bottom of the flashing lap the next shingle down. Center the flashing over the conduit centerline. Measure the length of conduit needed to put the top of the conduit at least 13 feet above grade and about 3 feet above the roof. Cut the conduit and ream or file the cut end. Then pass it down, threaded end first, through the holes and hangers, and thread

it tightly into the meter-socket hub. Install and tighten the hanger screws and then nail the flashing in place. Slip the boot over the mast and push it down as far as it will go. Install wire hangers as previously described. Slip the service-entrance head onto the mast and lock it in place with the setscrews or clamp.

Installing the Conductors

For connection to the service drop, 3 feet of conductor must extend out of the service head.

Add this to the distance measured from the top of the mast, through the meter socket, to the main lugs of the panel. Add 3 feet for good measure, and buy two pieces of black wire and one piece of white wire cut to this length.

Take the cap off the service head. Tape the three wires together at one end and push them up the mast until about 3 feet are sticking out of the top of the mast. Use a fish tape to pull the taped wires into the conduit when there are bends in the service conduit. Go to the top of the mast and bend the wires over and temporarily tape them to the mast. Return to the socket and cut the black wires—but not the white—so they'll fit into the top set of lugs in the meter socket. Strip just enough insulation from each wire so that the bare conductors will fit entirely into the lugs. Insert the wires and tighten the lugs. After tightening, work the wires back and forth a bit and then tighten the lugs some more. The white wire fits into a "lay-in" neutral lug in the center of the socket. Remove a band of insulation from the wire that is just a bit wider than the lug is long, but do not cut the wire.

Tape the two remaining black wires, which were cut off the wires pushed up the service conduit, to the free end of the white wire and push all three through the nipple and into the panel. Strip the black wires and connect them to the bottom set of lugs in the meter socket. Then lay the white wire in its lug and tighten it. Push

all the excess wire out of the socket enclosure into the panel. Route the wires into place in the panel, the black wires to the main breaker lugs, the white to the neutral bus. Strip and connect them, working them back and forth after tightening them, and then tightening them some more.

The cap of the service head has a plastic-like insert with knockouts in it. Break out three of these that are as far from one another as possible. Insert the white wire in the center opening and the black wires in the remaining openings. Slide the cap up the wires and over the service head, and tighten the two screws that hold it in place.

The service conduit must be both weathertight and arranged to drain. The service head is designed to be weathertight, and a means of conduit drainage is accomplished by providing a ¼-inch weep hole in the bottom of the meter socket, or SLB (service el conduit body) if one is used. Weathertight fittings are necessary when EMT is used as service conduit. There's always the possibility that condensation may occur in a conduit that runs between the warm indoors and cold outdoors, such as where a service conduit enters a house. Pack sealing compound into the nipple, or SLB, to prevent the movement of air through the conduit.

Underground Service

Everything after the meter socket is the same for an underground service as for an overhead service, but installation of the conductors up to the meter socket is different. In many

areas the power company brings the service-lateral cable across your property and connects it to the lugs at the top of the meter socket. Their trenching and cable installation is done mechanically and leaves almost no damage in the yard. It's your responsibility to provide a piece of conduit for protecting the cable where it emerges from the ground. This conduit must be long enough to reach from the bottom of the meter socket to a point 24 inches below grade. Typically, 2-inch Schedule 80 PVC conduit is used here, but rigid metal conduit or IMC (intermediate metal conduit) may also be used. The PVC should have a threaded adapter solvent-cemented to both ends. A non-metallic bushing is screwed onto the end that terminates in the ground, and a locknut and nonmetallic bushing must be left on the top end for attachment to the meter socket. This assembly is usually left next to the meter socket for eventual installation by power company employees. When metal conduit is used to provide this protection it, too, must have a nonmetallic bushing installed on the trench end and a locknut and an insulated grounding bushing at the meter end.

Consider renting a trencher if it's necessary for you to do your own trenching in order to reach the power company's mains or transformer. Contact the power company before beginning. A representative will ask where you will be digging, and will tell you how far to go and what wiring method to use. Also, clear where you intend to dig with other utilities such as the gas company, the

telephone company, and the water department, in case they have pipes or wires buried in your pathway.

Service-lateral trenches dug across residential properties need be only 6 inches deep for rigid metal conduit and IMC, 18 inches deep for PVC conduit, and 24 inches deep for direct-burial conductors and cables. All of these wiring methods require an 18-inch-deep trench where they pass under a residential driveway. After the power company and other utilities have approved your trench route, roll up the sod, set it aside, and start digging. Keep the trench as straight as possible, maintain a uniform depth, and place the excavated dirt within a few feet of the trench. Make the trench the width of a shovel except at the ends, where it is best to make it a bit wider and deeper.

There's nothing particularly difficult about installing conduit in a trench. The most important thing to remember is that no rocks or dirt can be present in the conduit. This means you should look through each length before installing it; and, when installing, be certain to keep the open end up out of the trench. Let the open end rest on a board or shovel laid across the trench as you go along. If for any reason an open conduit end must be left in the trench, cover the end with plastic and tape it up. You can keep groundwater from accumulating in your conduit by installing a drain tee at the low point of the conduit. A 4-inch-long nipple should be threaded into the tee and pointed down into a 1-foot-deep hole that has been filled with gravel.

THE METER, MAIN CIRCUIT BREAKER, AND SERVICE-ENTRANCE PANEL

The installation examples given in this section are for 100-amp services. The techniques used for larger services are basically the same except that larger-sized materials are used. When you select your service size, be sure to allow room for future expansion or amenities such as spas and pools.

The Meter and Main Circuit Breaker

Meter sockets are selected for the size of the service. The two most common sizes are 100 and 200 amp. Larger services than these require special metering arrangements that use current transformers to reduce the current in the metering circuit to a low value that's proportionate to the actual amount of current being delivered to the service. The power company can tell you the type and size meter socket you'll need for your service. Local electrical suppliers can give you this information as well.

When the service-entrance panel is installed indoors and the main circuit breaker is installed outdoors, the breaker is sized to protect the service conductors and to limit the current to the mains to a safe value. A 100-amp service is protected by a 100-amp circuit breaker, and so forth. Some authorities allow a main fused switch to be used in place of a circuit breaker. Often, the meter socket and main disconnect are mounted in a common enclosure. If they're separate, they should be mounted side by side.

All equipment enclosures used outdoors must be Type 3R, which means that they are constructed to provide an adequate degree of protection against falling rain and sleet and external ice formation. Outdoor enclosures are easily recognized because their tops extend beyond their front and sides to deflect rain. Weathertight hubs must be used when conduit enters the top or sides of the enclosure. The circuit-breaker enclosures and fused switches to be used as service disconnects must be marked as being suitable for use as service equipment.

All electrical equipment must be firmly secured to the surface on which it is mounted. It's unwise to mount equipment on the side of a building and depend only on the siding material for support, unless the siding is wood. Screws that pass through other types of siding into wood sheathing provide excellent mounting support. Sometimes it's necessary to provide blocking inside a wall, or to cut a board into the siding and nail it to the studs to provide sufficient support. When mounting equipment on masonry walls, use masonry anchors and screws, not screws and wooden plugs driven into holes in the wall.

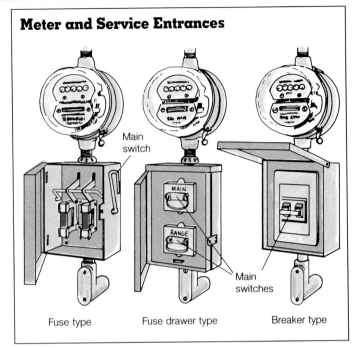

Meter and Service Entrances

Main switch

Main switches

Fuse type   Fuse drawer type   Breaker type

Outdoor enclosures have to be mounted in such a way that moisture can't accumulate between the back of the enclosure and the surface on which it is mounted. The mounting holes in equipment enclosures are drilled through bumps that extend outward from the back of the enclosure. These bumps are there so that the enclosure won't twist if mounted on an irregular surface and so the back of the enclosure won't contact the surface on which it is mounted, which would allow moisture to accumulate. The top and sides of the enclosure must be caulked at the wall.

If the meter and main breaker are in a common enclosure, the black wires entering from the service conduit connect to the top lugs on the socket. The connections between the meter's bottom lugs and the line side of the main breaker are factory-connected. The white wire connects to the neutral lug in the circuit break half of the enclosure. Two blocks are then used to connect the load side of the main breaker to the panel main lugs, and a white wire is used to connect the neutral lug to the panel's neutral bus.

When the socket and main breaker are in separate enclosures they should be mounted side by side. The top of the socket is fed as in the previous example, and field-installed wires complete the connections between the bottom of the socket and the line side of the main breaker. The white connects to a lay-in lug in the socket and continues to the neutral lug in the breaker enclosure. Final connections to the breaker panel are the same as in the previous example.

Specific requirements vary from locality to locality. Always contact your electrical-inspection department regarding particulars.

Wiring a Meter Socket

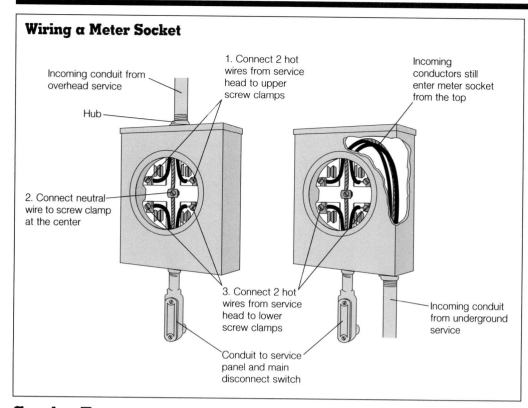

Incoming conduit from overhead service

Hub

1. Connect 2 hot wires from service head to upper screw clamps

Incoming conductors still enter meter socket from the top

2. Connect neutral wire to screw clamp at the center

3. Connect 2 hot wires from service head to lower screw clamps

Conduit to service panel and main disconnect switch

Incoming conduit from underground service

Sizing and Installing the Service Conductors

The size of the service conductors is determined by the size of the service panel you've chosen. The nominal size of copper service wires used with 100-amp services is No. 4 with a temperature rating of 194° F (90° C). No. 1 is used with 150-amp services, and 2/0 (two-ought) with 200-amp services. If you use aluminum wire, larger sizes are needed to provide the required ampacities.

Service-Entrance Panel

The minimum allowable service size for a house is determined by using one of the calculation methods allowed by Article 220 of the Code (see page 20). These calculations take into consideration the size of the house, the required 120-volt circuits, and the 240-volt branch circuits that supply fixed loads such as ranges, water heaters, and clothes dryers. The service current and the number of branch circuits needed for general lighting, which includes the permanent lighting and most of the convenience outlets in the house, also are determined by this process. To calculate the number of fuses or circuit-breaker spaces needed in addition to those for the general lighting load, add up the required small-appliance circuits,

the laundry circuit, plus any other 120-volt individual branch circuits, such as for the dishwasher and the food disposer. Each of these circuits requires one circuit-breaker space or fuse. In addition, two spaces are needed for each of the 240-volt individual branch circuits. These are the circuits that feed the range, water heater, and clothes dryer. Add them all up to know how many circuit breakers the panel will need.

If your initial service-current calculation is just a little less than a standard service size, 100, 150, or 200 amp, you should think about going to a larger-than-required service to allow for future expansion. Also, consider the number of circuit-breaker spaces your planning calls for. One-hundred-amp panels are generally limited to 20 spaces, so more spaces means a larger panel and associated service wiring.

Installation Techniques

Panels should be mounted on wood surfaces for soundness and ease of fastening. Often, a large piece of ¾-inch plywood is fastened to a basement masonry wall with anchors or concrete nails, then the panel is secured to the plywood with wood screws. The plywood should be wider than the panel and should extend at least 1 foot above and, if necessary, 1 foot below the panel in order to provide a surface to anchor the cables and conduits that run to the panel. Remember to provide adequate space behind the panel for air circulation. If the bumps alone aren't sufficient, run the mounting screws

through ¼-inch nuts held between the panel and the board, which will guarantee more standoff from the wall. A 1¼-inch conduit is run from the bottom of the meter socket into an SLB at the rim joist. A hole would be sawed at the rim joist where a conduit elbow would pass through it to connect the SLB to the panel.

Studs provide excellent mounting for flush-mounted installations. Provide blocking above and, if needed, below the panel for cable anchoring.

A panel in a garage is commonly mounted between studs, with the top edge about 5 feet above the floor. Most modern panels are about 14⅜ inches wide, so they fit perfectly between studs that are 16 inches on center and can be secured in place with two screws through each side of the panel. Begin by removing a 1¼-inch top knockout from the rear of the panel and tracing its outline on the wall in the desired position. Using a hole saw or dial saw, cut a 2½-inch hole through the wall at the center of the traced outline.

Go to the exterior side of the wall and open a 1¼-inch bottom knockout in the back of the meter socket. Thread locknuts completely onto both ends of a 1¼-inch by 2½-inch conduit nipple and insert one end into the knockout. Thread another locknut and then an insulated grounding bushing tightly onto the nipple on the inside of the socket enclosure. Push the nipple into the center of the hole in the wall, square up the enclosure and secure it in place with wood screws.

Back in the garage, pack the hole in the wall with caulk and slip the panel over the nipple. Thread a locknut and then an insulated grounding bushing tightly onto the nipple, plumb the panel, and fasten it in place with wood screws.

Outdoor installations must be caulked where the panel top and sides meet the wall, but you can leave the bottom open.

Replacing an Existing Service Panel

Replacing an old fuse panel with a new, larger circuit-breaker panel is usually best left to a licensed electrician. Many municipalities also require the addition of small-appliance circuits to the kitchen when the service in an older home is upgraded. No two fuse-to-circuit-breaker changeovers are alike.

If your main disconnect is in the panel, the power company must remove your meter to disconnect the power to your panel before you begin. Then, after your changeover is complete and has passed inspection, call the power company again to reinstall the meter.

Install any required additional circuits and drive an 8-foot ground rod near the point where the service conductors enter the house, if one isn't already there, before beginning the actual panel changeover. Then have the utility company disconnect your power at the pole and remove your meter. If the existing service conduit is 1 inch or larger, it won't have to be changed; if it is less than 1 inch, remove it.

Identify the branch-circuit wires with color-identification tape or wire markers. Then disconnect the wiring inside the panel, swing the cables or conduits out of the panel, and remove the panel itself. Install the new service conduit, if necessary, and fit the new panel to the service conduit and secure it in place. Pull in and connect the new service wires, and also install a grounding-electrode conductor from the ground rod to the neutral bus in the service panel. When the service installation is complete, call the building-inspection department that issued your permit to schedule an inspection. After the installation is approved,

Wiring a Service Panel

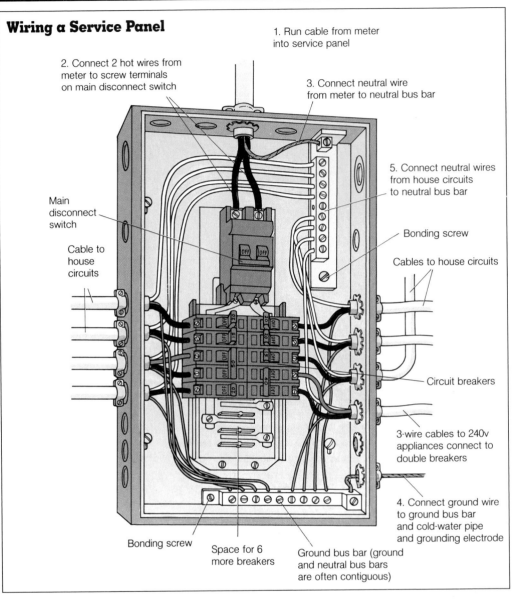

1. Run cable from meter into service panel

2. Connect 2 hot wires from meter to screw terminals on main disconnect switch

3. Connect neutral wire from meter to neutral bus bar

5. Connect neutral wires from house circuits to neutral bus bar

Bonding screw

Cables to house circuits

Circuit breakers

3-wire cables to 240v appliances connect to double breakers

4. Connect ground wire to ground bus bar and cold-water pipe and grounding electrode

Main disconnect switch

Cable to house circuits

Bonding screw

Space for 6 more breakers

Ground bus bar (ground and neutral bus bars are often contiguous)

you or the inspector (depending on local practice) call the electric utility company to request an immediate hookup.

Reconnect the branch-circuit wiring if it wasn't required for your final inspection. As you work make a note of which circuit goes to which breaker so that you can properly fill out the circuit directory for the panel when you're through. Often, because the new panel is physically larger than the old one, some of the wires won't reach their circuit breakers. Splice extensions to the old wires inside the panel when this happens.

Types of Overcurrent Devices

Unlike fuses, which use nationally standardized thread designs, circuit breakers are designed by panel manufacturers to be used in their panels only and are generally not interchangeable.

Breakers used in residential panels are either single-pole, for 120-volt circuits, or two-pole (double-pole), for 240-volt circuits. Common ampere ratings for single-pole breakers found in the home are 15 and 20 amp. Some common two-pole ratings are 15, 20, 30, 40, and 60 amp, plus 100, 150, and 200 amp main circuit breakers.

Full-width single-pole circuit breakers are about 1 inch wide, but some manufacturers have made ½-inch-wide breakers. "Piggyback" and "twin" circuit breakers are available for some panels to allow additional circuits to be powered from an already full panel. GFCI circuit breakers are also available. Circuit breakers make an electrical connection

to the main bus of the panel as well as a mechanical nonconducting attachment to a rail or bar which holds them firmly in place.

Buses

Competitively priced panels use aluminum main buses. These are fine in most parts of the country, but premium copper bus panels should be used in coastal areas where salt air can have a damaging effect on aluminum.

The neutral bus in a service is bonded to its enclosure so all branch-circuit grounding wires as well as all neutral conductors are connected to it. A subpanel differs from a service panel in that the neutral bus of a subpanel is not bonded to its enclosure, and a separate grounding bus must be provided for the branch-circuit cables being powered by it. A grounding bus, or grounding bar kit, can be installed for convenience in a service panel, but it is not required.

Installing Circuit Wires in the Service-Entrance Panel

Service panels are much larger now than they were in the past because of recent *NEC* changes in wire-bending space requirements. The increased space available means that panel wiring can be more systematic and easier to trace than it was in the past. It's best to strip the cable jacket to within ½ inch of where the cable enters the enclosure; then route the individual conductors to the rear and side of the panel to a point

Fuses and Circuit Breakers

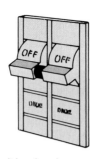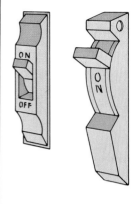

Three styles of circuit breakers by different manufacturers. The double breaker (at right) has a bar connecting the 2 handles and is used to protect 240v circuits.

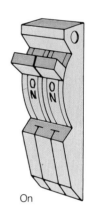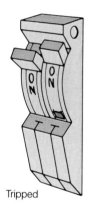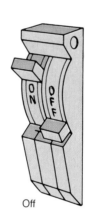

On   Tripped   Off

Some breakers go to the *off* position when tripped. Some go halfway to *off* or (as shown above) to a *tripped* position. These must be switched to *off* then *on* to restore service.

You can replace your screw-in fuses with this type of breaker (at the same rating). An overload will cause the button to pop out, exposing a colored band. To restore service, just push in the button.

level with where they'll attach to a circuit breaker or the neutral bus. At this point bend the wire so that it points straight out at a distance out from the back of the panel that is equal to the distance out to the terminal. Bend the wire toward the

terminal, cut it to fit, strip, and connect it.

Be sure to accurately fill in the circuit directory for the panel. It can save a lot of confusion and uncertainty when there is trouble.

Circuit Breakers

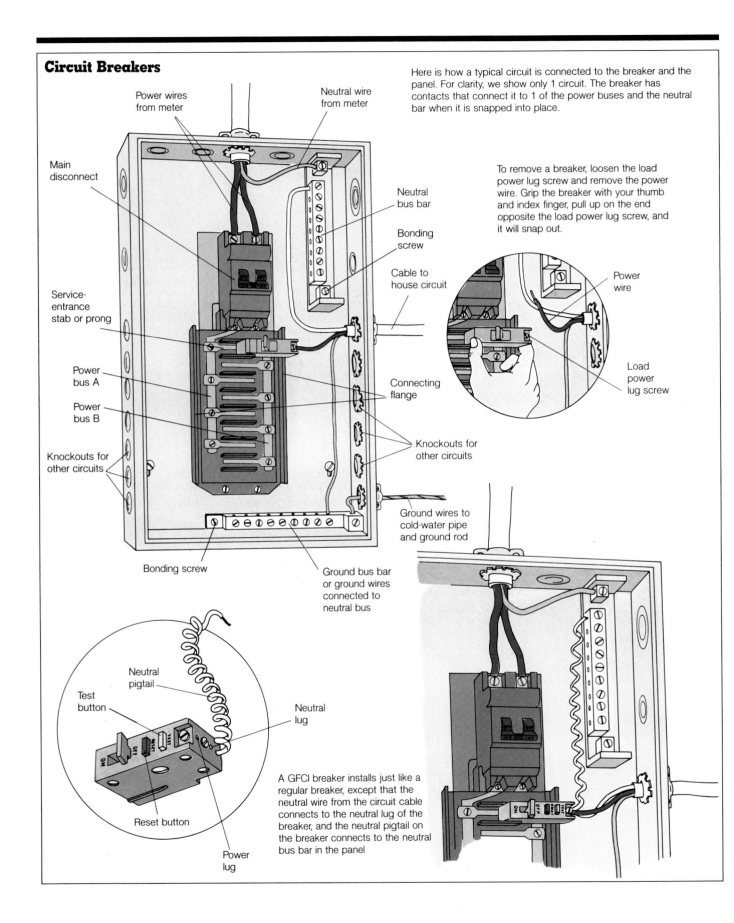

Power wires from meter

Neutral wire from meter

Here is how a typical circuit is connected to the breaker and the panel. For clarity, we show only 1 circuit. The breaker has contacts that connect it to 1 of the power buses and the neutral bar when it is snapped into place.

Main disconnect

Neutral bus bar

Bonding screw

Cable to house circuit

To remove a breaker, loosen the load power lug screw and remove the power wire. Grip the breaker with your thumb and index finger, pull up on the end opposite the load power lug screw, and it will snap out.

Power wire

Load power lug screw

Service-entrance stab or prong

Power bus A

Power bus B

Connecting flange

Knockouts for other circuits

Knockouts for other circuits

Bonding screw

Ground bus bar or ground wires connected to neutral bus

Ground wires to cold-water pipe and ground rod

Neutral pigtail

Test button

Neutral lug

Reset button

Power lug

A GFCI breaker installs just like a regular breaker, except that the neutral wire from the circuit cable connects to the neutral lug of the breaker, and the neutral pigtail on the breaker connects to the neutral bus bar in the panel

THE GROUNDING SYSTEM

All the equipment grounding in the house is connected to the neutral bus in the service panel. The neutral service conductor, which is grounded at the distribution system, connects to the neutral bus along with the grounding-electrode conductors that connect the bus to a driven ground rod and the metal water piping of the house.

The Ground Rod

This ½-inch-diameter copper-clad rod is driven 8 feet straight into the ground at a point outside the house, usually near where the service conductors enter the house. The rod can be driven at an angle not to exceed 45 degrees from the vertical, or laid in a trench 2½ feet deep, if solid rock is encountered less than 8 feet down. No part of the rod can be above ground level once the installation is complete. In place of ground rods, ¾-inch galvanized pipe, rigid metal conduit, or plate electrodes can be used; aluminum grounding electrodes cannot.

The Grounding-Electrode Conductor

A wire called the grounding-electrode conductor connects the ground rod to the neutral bus in the panel. It connects to one of the larger terminals in the neutral bus, and a ground clamp connects it to the ground rod. Typically, this conductor is bare copper and is sized according to the service: No. 8 copper for 100-amp services, No. 6 for 150, and No. 4 for 200. The grounding-electrode conductor must never be spliced.

An additional grounding-electrode conductor connects the interior metal water piping of the house to the neutral bus. This connection can be made to a cold-water pipe at any convenient point, again using a ground clamp at the pipe and one of the larger terminals in the neutral bus. Aluminum wire is often used here because the wire doesn't contact the earth. Use No. 6, No. 4, and No. 2 aluminum wire, respectively, with 100-, 150-, and 200-amp services.

There may be additional local requirements, for example, shielding the ground wire in conduit where it would otherwise be exposed to damage, such as in a garage or driveway where a car could back into it.

Bonding

All the metal conduits and enclosures that are part of the service must be bonded (electrically connected) together. Threaded conduit connections and hubs on enclosures and conduit bodies provide adequate bonding. Where conduit is connected to an enclosure with locknuts and bushings, the bushing should be a metal grounding bushing with an insulating insert. This type of bushing has an attached lug that allows a bonding jumper to connect to it. The other end of the bonding jumper attaches to a terminal inside the enclosure.

A bonding jumper must be installed around the water meter. Ground clamps and a jumper that's long enough to permit removal of the meter without having to remove the jumper are used to meet this requirement. Some municipalities also require bonding jumpers around water heaters and spas. The required bonding jumpers must be the same size as the grounding-electrode conductor of the system.

Grounding at a subpanel is a bit different than grounding at the service. The neutral bus is not bonded to the subpanel enclosure. A separate grounding bus is used to connect the grounding conductors of the cables to the ground wire of the branch feed.

Sizes of Grounding Electrode Conductors

Size of Largest Service-Entrance Conductor		Size of Grounding Electrode Conductor	
Copper	Aluminum or Copper-Clad Aluminum	Copper	Aluminum or Copper-Clad Aluminum
2 or smaller	1/0 or smaller	8	6
1 or 1/0	2/0 or 3/0	6	4
2/0 or 3/0	4/0 or 250 MCM	4	2
Over 3/0 through 350 MCM	Over 250 MCM through 500 MCM	2	1/0
Over 350 MCM through 600 MCM	Over 500 MCM through 900 MCM	1/0	3/0
Over 600 MCM through 1100 MCM	Over 900 MCM through 1750 MCM	2/0	4/0
Over 1100 MCM	Over 1750 MCM	3/0	250 MCM

Reprinted with permission from NFPA 70-1990, the *National Electrical Code*, Copyright © 1989, National Fire Protection Association, Quincy, MA 02269. This reprinted material is not the complete and official position of the National Fire Protection Association on the referenced subject, which is represented only by the standard in its entirety.

 INISH WIRING

Up to this point you have accomplished quite a lot. You have mapped out your circuits, installed the switch and receptacle boxes in the appropriate locations, run the correct cables into the boxes, and installed the service panel. After these installations are inspected and approved, and the walls and ceilings are finished, you are ready to make the final connections and turn on your electrical service. This hook-up step, which is called finish wiring, involves three specific stages. First you connect the cable in each box to the appropriate receptacles, switches, and light fixtures. Next you make the hard-wired connections at all the fixed appliances and equipment. Finally you make the branch-circuit connections in the service panel. Installing cover plates at the switches and receptacles completes the job. Finish wiring is by no means the most complicated part of wiring a house, but it still requires care and attention to detail.

There can be more to finish wiring than attaching conduits to receptacles or switches. The kitchen or laundry room can require an array of hookups.

All finish work must be done with care. The wires must be connected to the fixture properly, of course, and the fixture itself must be plumb and level and fit flush against the wall or ceiling with no gaps showing.

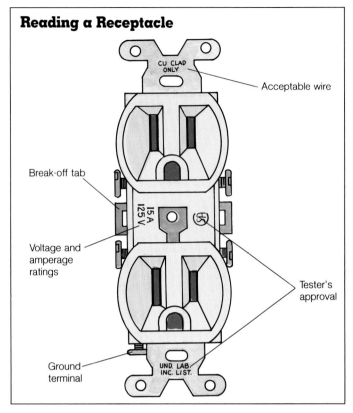

Reading a Receptacle

CU CLAD ONLY

Acceptable wire

Break-off tab

15A 125V

Voltage and amperage ratings

Tester's approval

Ground terminal

UND. LAB. INC. LIST.

Installing Receptacles

Receptacles should be held plumb while you tighten their mounting screws. After tightening, step back and check your work. If it is necessary to adjust, or "tune," it, hold a screwdriver tip against the top or bottom of the device strap near the mounting screw and gently tap the screwdriver with pliers until the device is plumb.

If the electrical boxes were installed properly, the finish wall and ceiling materials should be flush with the front edges of the boxes. When they are, installing the devices is easy, and their finished appearance looks smooth and professional because the plate fits the device with no unsightly gaps around the edges. A box that extends only slightly beyond the wall usually isn't a real problem because the shape and flexibility of the plate will adjust for it. If the wall surface extends beyond the edge of the electrical box, the plaster ears on the device strap will keep it from being drawn into the wall when it is attached to the box. Occasionally, especially in existing plaster walls, the opening around the box is too large

for the plaster ears to be effective. One remedy for this is to space the device away from the box and flush with the wall surface with washers or nuts that the mounting screws pass through.

Installing 120-Volt Receptacles

When wiring a 120-volt duplex receptacle, connect the white wire to the silver-colored screw, the black (or possibly red) wire to the brass-colored screw, and the green (or bare) wire to the green screw. All the connections are made by first stripping 1 inch of insulation from the ends of the wire; then use needle nose pliers to bend the end of the wire into a loop, and slip the loop over the body of the screw in such a way that tightening the screw will tend to close the loop. Close the loop with the pliers and tighten the screw. Note that back-wired receptacles don't require this wire looping. Instead, strip the wire to the length indicated by the strip gauge for the device. Then, depending on the type of back-wiring, either simply

push the stripped wires into their appropriate holes, or push the wires into the proper openings and tighten the terminal screws. Tug on each wire to check for a tight connection.

If you should need to remove the wires from a back-wired push terminal, there is a release slot for this purpose. You can simply insert a small screwdriver into this slot to release the wire.

If you are installing the two receptacles side by side, use the second set of terminal screws on the first receptacle to refeed the second receptacle. Never use the second set of screws to feed additional receptacles in other boxes. Instead, use pigtails connected to the incoming and outgoing cables to connect to the receptacle. Some local codes may require pigtail connections even for side-by-side receptacles.

Installing 240-Volt Receptacles

Some 240-volt equipment is always hard-wired to its source of power. Two examples are water heaters and central air-conditioning equipment. There is no reason to cord-connect these major appliances because they don't have to be moved in order to be serviced. On the other hand, ranges, cooktops, and wall-mounted ovens can be hard-wired, but it usually makes more sense to use cords and plugs with them for ease

Wiring Receptacles

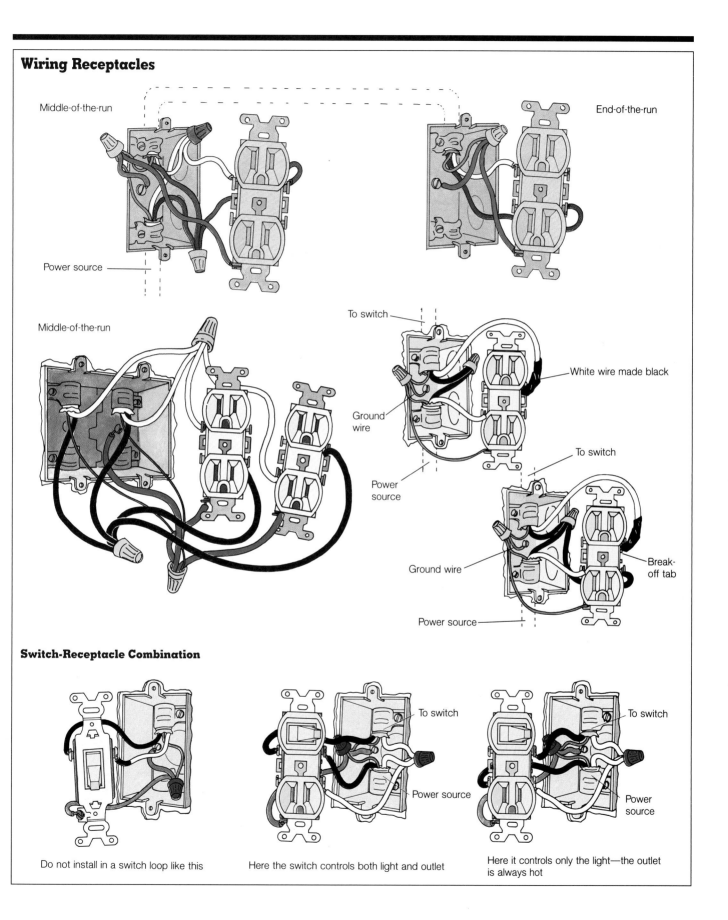

Middle-of-the-run

End-of-the-run

Power source

Middle-of-the-run

To switch

White wire made black

Ground wire

To switch

Power source

Ground wire

Break-off tab

Power source

Switch-Receptacle Combination

To switch

Power source

To switch

Power source

Do not install in a switch loop like this

Here the switch controls both light and outlet

Here it controls only the light—the outlet is always hot

Color Coding of Wires and Terminal Screws for Switches, Receptacles, and Light Fixtures

Function of Wire	Color of Wire	Color of Terminal Screw
Neutral wire	White	Silver or white
Hot wire	Black	Brass
Hot wire	Red	Brass
Grounding wire	Green	Green
Grounding wire	Bare wire	Electrical box ground

Note: A white wire used as a hot wire, such as when using Type NM cable for a switch loop, must be made black at both ends.

Connecting Wires to Switches

1. Push in the push-in terminal

2. On the gauge on switch, measure length of exposed wire

3. Push exposed wire into opening in terminal

A Note on Wire Color Coding

The neutral wire is white and always connects directly to the light fixture and never to the switch. Otherwise, the light could still work but every time the switch was turned on there would be a dangerous potential for shock at the light fixture because the neutral wire could no longer provide a safe, unbroken path back to the source.

Switches should always be connected to the hot wire, not the neutral. In most cases the hot wire is either black or red. But if you are using nonmetallic cable for the switch loop between the switch and a light fixture fed directly by the source, one of the wires in that cable will be white, even though it will function as a hot conductor. This hot conductor is called a "white made black" wire, and it should be marked at both ends with black electrical tape, black paint, or other black marking to indicate its status as a hot conductor. This wire and the true black wire are interchangeable.

At one end of the cable both wires are connected to the switch; at the other end one wire is connected to the light fixture at the brass-colored screw or black pigtail wire, and the other wire is connected to the hot conductor from the source with a wire nut or other approved electrical splice.

of disconnecting when they require servicing. Clothes dryers are cord-connected for the same reason.

Wiring for a 240-volt receptacle begins at the service panel and terminates in an electrical box located on the wall at the appliance location. The run can be spliced, inside a junction box, but must not have any branches or other receptacles on it. The wire must be sized for the amperage of the appliance (e.g., No. 10 wire and a 30-amp circuit breaker for a 30-amp dryer; No. 6 wire and a 50-amp breaker for a 50-amp range). Not all 240-volt receptacles are identical. Choose a receptacle which is sized for the amperage of the appliance and wire. It must also match the plug configuration on the appliance cord.

Install the receptacle by attaching the feed wires of the cable to the screw terminals on the back of the receptacle: Always connect the hot wires to the brass terminals, and, depending on the equipment, the green, white, or bare wire to the remaining terminal. Note that wiring for some 240-volt appliances, such as water heaters, does not require a separate ground wire insofar as the neutral wire serves that function.

Installing Switched Receptacles

The *NEC* requirement for wall-switch controlled lighting in all of the habitable rooms in the house is usually met by switching the top half of one receptacle in each of the bedrooms and the living room. The other half of the receptacle is powered all the time. When this is done, the receptacle halves are electrically isolated from one another by twisting out the break-off tab that ties the two brass-colored screws together, and connecting the hot wire to the bottom brass-colored screw, and the wire coming from the switch to the top one. The neutral wire connects to the silver-colored screw and serves both halves of the receptacle.

Combination switches and receptacles look essentially like ordinary duplex receptacles except that the top half of the device contains a toggle switch that operates horizontally. They provide a means of switching a circuit and having 120-volt power available at a single one-gang switch box. Some of these devices are furnished with a built-in neon glow lamp that indicates when the switch is on. A variation of this combination device provides a three-way switch with a grounding receptacle.

Wiring a Single-Pole Switch

Source Through Switch

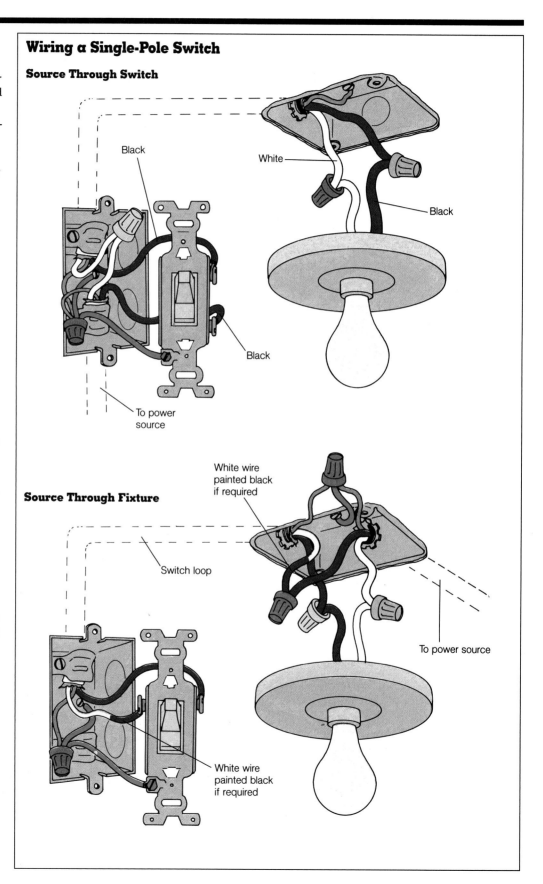

Black

White

Black

Black

To power source

Source Through Fixture

White wire painted black if required

Switch loop

To power source

White wire painted black if required

Installing or Replacing a Three-way Switch

Power Coming from Source

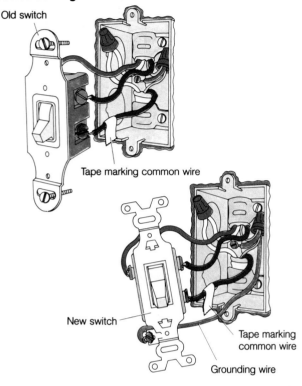

Old switch

Tape marking common wire

New switch

Tape marking common wire

Grounding wire

Power Coming from Fixture

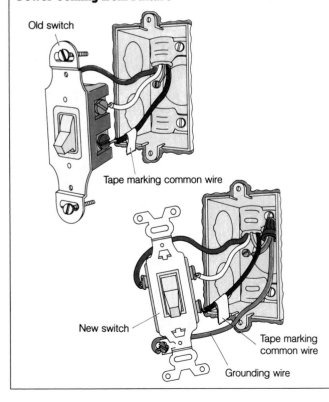

Old switch

Tape marking common wire

New switch

Tape marking common wire

Grounding wire

Installing GFCIs

GFCI duplex receptacles are found in nearly all new houses. GFCI circuit breakers combined with ordinary receptacles provide the same protection; but GFCI receptacles have the added convenience of the reset and test capabilities available at the receptacle rather than at the service panel. It's common practice in kitchen wiring to run one of the two required small-appliance circuits directly to a GFCI located near the sink, and then to feed any other receptacles in the kitchen that require ground-fault protection from that GFCI.

GFCIs are physically larger than ordinary duplex receptacles, which means they can't always directly replace an ordinary receptacle. They'll fit in most switch boxes, provided the box isn't loaded with wires, but they don't always fit in houses that are wired with square boxes and mud rings. They can be used to advantage as replacements for ordinary receptacles in older houses with ungrounded branch-circuit wiring. GFCIs offer excellent shock protection even when there's not a grounding conductor connected to them.

Installing Switches

Most houses will use a variety of switches that may include two-way, three-way, four-way, and dimmers. The wiring is slightly different for each type.

Two-way Switches

The most common, and easiest to understand, switch is the ordinary on-off single-pole switch known as a two-way. It has two screws to which the circuit conductors are connected, and it makes no difference which conductor is connected to which screw. The single-pole switch isn't used in combination with other switches, and like most switches sold today it has a ground terminal screw on its strap.

Three-way Switches

Switches used in pairs to control lighting from two different points are called three-way switches. They're characterized by three circuit-conductor screws, two brass-colored and one bronze-colored, and the absence of *on* and *off* labels on the switch handle. The bronze-colored screw is connected internally to the shunt, or common, connection. Moving the switch up connects the common with one of the brass-colored screws, moving it down connects it to the other.

Three-way switches can be connected in a number of ways. The simplest system involves connecting the "hot" lead to the shunt of one switch (the dark or bronze-colored screw marked as "common") and connecting a pair of wires between the brass-colored terminals of one switch and those

Wiring Switches and Receptacles

Source Through Switch

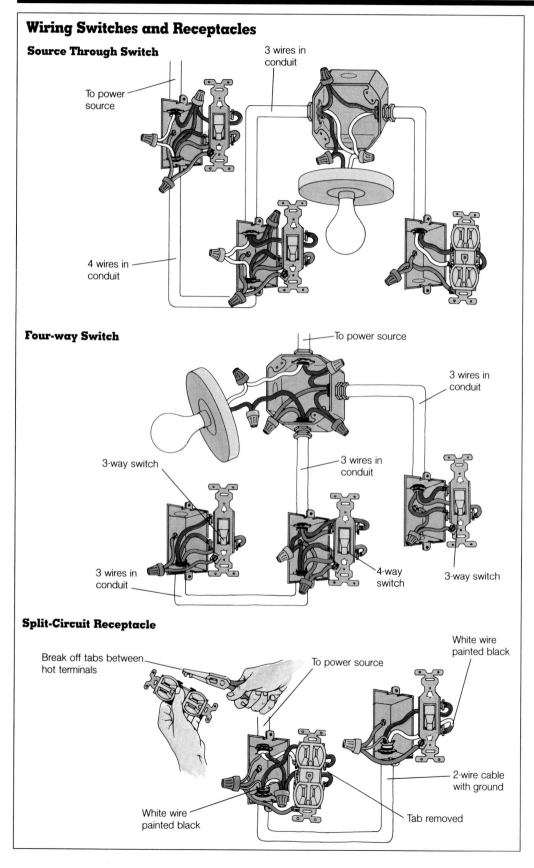

To power source

3 wires in conduit

4 wires in conduit

Four-way Switch

To power source

3 wires in conduit

3-way switch

3 wires in conduit

3 wires in conduit

3 wires in conduit

4-way switch

3-way switch

Split-Circuit Receptacle

Break off tabs between hot terminals

To power source

White wire painted black

White wire painted black

2-wire cable with ground

Tab removed

of the second switch. These wires are called travelers. Finally, the shunt terminal of the second switch is connected to the light that's being controlled, along with, of course, the neutral wire.

Another common three-way arrangement employs switch loops. Often, branch-circuit wiring is run directly to a ceiling box where a light is to be installed. In this case, three wires are run to each switch location. When cable is used, you connect the black wire to the bronze-colored screws on both switches, and the red and white wires to the brass-colored screws. Back at the ceiling box, splice the three-conductor cables red to red, and white to white. Connect the hot wire to either black wire, and the other black wire to the brass-colored screw or colored wire at the fixture; it makes no difference which goes where. Then connect the neutral wire to the fixture.

Four-way Switches

For additional points of lighting control between pairs of three-way switches, four-way switches are available. Any number of four-ways can be used, but they must be installed between a pair of three-way switches. Four-way switches can be identified by

Installing a Dimmer Switch

Two-way Dimmer

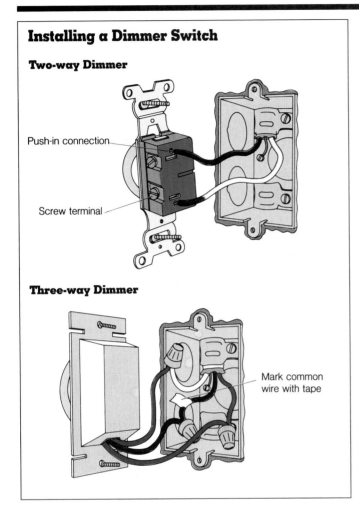

Push-in connection

Screw terminal

Three-way Dimmer

Mark common wire with tape

their four circuit-conductor terminal screws and the absence of *on* and *off* on their operating handles.

Dimmers

Dimmers are available in both single-pole and three-way forms. The dimmers that are commonly available are exclusively for incandescent lights, and should be used only to control fixed lighting, not the switched receptacles found in most homes. Fluorescent-lamp dimming requires a special

dimming ballast and an auxiliary control that's matched to the ballast. These can be obtained from lighting wholesalers and seldom elsewhere.

When installing a dimmer, it's doubly essential that before you begin you turn off the power to the circuit on which you will be working. As always, you are ensuring your own safety, but you are also protecting the dimmer, which

would probably be damaged or destroyed if it were installed while the power was on. Replacing a single-pole switch with a dimmer is simply a wire-for-wire changeover. A three-way dimmer can replace one, but not both, of the three-way switches in a multiple switching arrangement.

Basic dimmers have either a knob that's turned clockwise to turn the lights on, or a knob that is pushed for *on* and pushed again for *off*; once on, the knobs of both types are rotated to adjust the level of lighting. A move up from this is the toggle control, which looks like an ordinary toggle switch and varies the lighting level as it's moved up and down. Slide dimmers are similar to the toggle type in operation, but the sliding operator doesn't extend outward. At the top of the dimmer line is the touch-control switch. This switch has a rectangular plate that fits almost flush to the wall. Touching the plate briefly and releasing turns the lights fully on; when the plate is touched a second time and held, the light level slowly varies until the plate is released. When the plate is touched and released again, the lights are turned off.

If you are counting on dimmers to save electricity, you need to investigate further before purchasing dimming mechanisms. Some will indeed reduce electrical use as lights

are dimmed, but others consume maximum electricity regardless of the lighting level by using a rheostat device which burns off any current not used by the light.

Replacing an Existing Switch

Always turn off the power when you replace a switch. Replacing a single-pole switch involves removing the two circuit wires plus the ground wire, if one is present, and reconnecting them to the new switch. Either circuit wire can go to either switch terminal because the switch will work correctly regardless of which goes where. Note that three- and four-way switches require greater care when replacing. It's best to make a diagram showing terminal and wire colors; you may need it to get the lights working properly again. If there are two or more wires of the same color at the switch, identify them with different numbers of narrow bands of electrical tape or some other handy means. Look at the screw colors carefully. Some are bronze-colored and some are brass-colored. After connecting, fold the wires carefully back into the box and fit the switch into the box before tightening the mounting screws to avoid damaging the wire.

INSTALLING LIGHT FIXTURES

Proper polarity must be observed when wiring light fixtures. Also, properly ground the fixture. It is standard practice, based on an NEC requirement, for manufacturers to connect the screw shells of their fixture lamp holders to the silver-colored screw, or white wire, which will be connected to the grounded neutral supply wire.

Surface-Mounted Fixtures

Most surface-mounted fixtures, such as those on walls and ceilings, mount on a strap or bracket that is first connected to the lighting outlet box. The straps are designed to fit the mounting dimensions of the standard outlet-box device, and then provide tapped mounting holes for attaching the fixture to it. Any combustible ceiling or wall material that's exposed between the canopy edge of the fixture and the edge of the outlet box must be covered with noncombustible material. The manufacturer usually provides a pad of fiberglass material about the size and shape of the canopy for this purpose.

Pendant-mounting refers to ceiling fixtures that are chain or stem-mounted beneath their canopies. When installing any ceiling fixture, the job goes much more easily with a helper to hold the fixture in position while you do the wiring.

Recessed Lighting

A typical recessed light assembly includes a lamp housing and a wiring connection box mounted on a frame that attaches to the house framing with a pair of built-in, adjustable bar hangers. In addition, the assembly includes a piece of trim that will give the fixture a finished look after the installation is complete. The assembly, or frame, is installed so that its lamp opening bottom will be flush with the finished ceiling. Branch-circuit wiring is run directly to the wiring connection box of the fixture. Lamping the fixture and installing the trim after the ceiling is finished completes the installation. The wiring in the wiring connection box can be accessed after installation by removing three or four screws that hold the housing to the frame, then pulling the housing down and out of the frame.

Some recessed fixtures are designed for use where they'll be in direct contact with thermal insulation. Others are for use in uninsulated ceilings or in ceilings where the insulation will be kept at least 3 inches from the fixture.

Standard fixtures are about 7½ inches deep, and special 5¾-inch fixtures are made for installation in 2 by 6 ceilings. Certain universal models have housings that can be adjusted from 5½ to 7½ inches in depth. Still others are available for installation in sloped ceilings. Many manufacturers make recessed fixtures for the remodel market. These fixtures can be completely installed from below the finished ceiling.

Utility-Area Lighting

Lighting fixtures in unfinished basements, attics, and garages are usually mounted directly onto surface-mounted outlet boxes. The porcelain lamp holders commonly found in these locations connect directly to the box with 8-32 screws. Fluorescent fixtures are also used in these areas, usually suspended on fixture chains, and cord- and plug-connected to their source of power.

Track Lighting

Track lighting provides an extraordinary degree of lighting flexibility. It allows light to be easily directed exactly where it's needed. The track used is an extruded aluminum channel that holds two electrical conductors, one in each side of the extrusion, that are mounted in continuous insulated supports. The inside surfaces of the conductors are bare. When a lamp fixture is snapped into the track, two terminals on the fixture contact the track conductors, and power is supplied to the fixture. Fixtures are also connected to ground through the track.

Track is usually available in 2-, 4-, and 8-foot lengths, and it can be cut to any desired length with a hacksaw. Sections are fastened together with special connectors when longer runs are needed. End caps provided by the manufacturer give the track a finished look. Track lighting is most commonly surface-mounted, but may also be pendant-mounted, and is often connected to branch-circuit wiring with a floating-feed device. All that's necessary for floating feed is that the track cross a lighting outlet box. At this point, a connector that is similar to the connectors on the individual fixtures is wired to the supply conductors and then snapped into the track in much the same way a fixture is. A canopy is installed that covers both the outlet box and the short portion of track that crosses the box. The track conductors become energized when power to the outlet box is turned on. Manufacturers of track lighting have engineered their systems so that proper electrical polarity is always maintained. The polarity lines inside the channel must be aligned when sections of track are installed, and the fixtures will snap into the track in only one way, which assures that proper polarity will be maintained at the lamp holder.

Fixtures are available in a number of styles for general illumination, accent lighting, and wall washing. Fixtures from different manufacturers are not interchangeable.

Installing Ceiling Fixtures

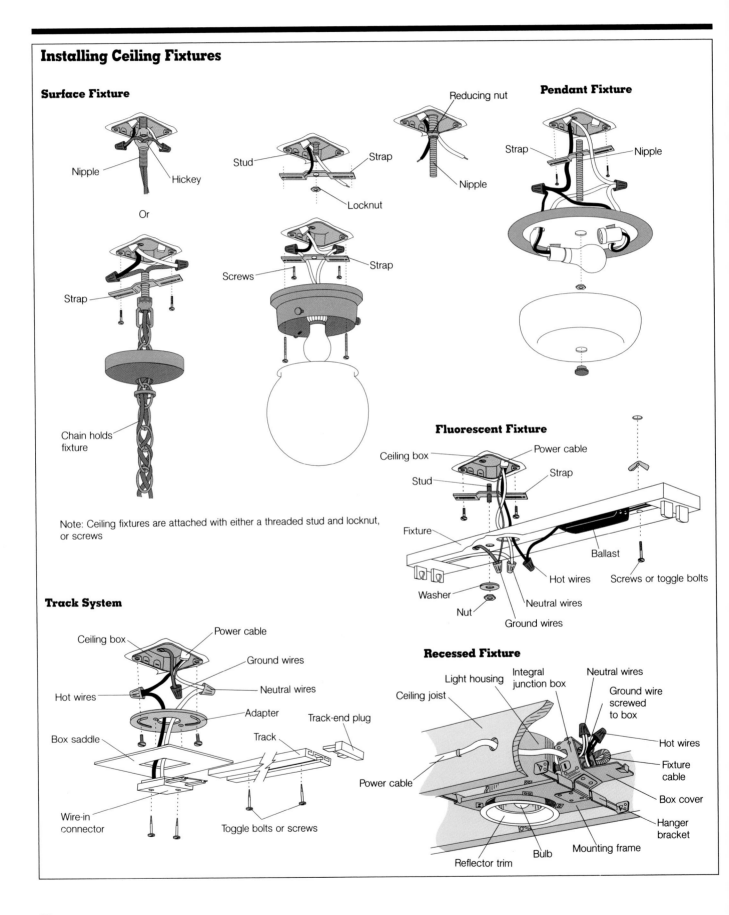

Surface Fixture

Nipple

Hickey

Or

Strap

Chain holds fixture

Stud

Strap

Locknut

Screws

Strap

Reducing nut

Nipple

Note: Ceiling fixtures are attached with either a threaded stud and locknut, or screws

Pendant Fixture

Strap

Nipple

Fluorescent Fixture

Ceiling box

Power cable

Stud

Strap

Fixture

Ballast

Screws or toggle bolts

Washer

Hot wires

Nut

Neutral wires

Ground wires

Track System

Ceiling box

Power cable

Ground wires

Hot wires

Neutral wires

Adapter

Track-end plug

Box saddle

Track

Wire-in connector

Toggle bolts or screws

Recessed Fixture

Light housing

Integral junction box

Neutral wires

Ceiling joist

Ground wire screwed to box

Hot wires

Fixture cable

Box cover

Power cable

Hanger bracket

Reflector trim

Bulb

Mounting frame

A variety of other options are available, but the two most commonly installed are ceiling fans and hard-wired smoke detectors. Ceiling fans require an approved mounting means. Smoke detectors should be wired into a frequently used branch circuit so that any loss of power will be quickly detected.

Ceiling Fans

Many owners of older homes have simply replaced an existing ceiling fixture with a fan and have had no trouble at all. However, this cannot be done in newer homes with their nonmetallic outlet boxes. By itself, this type of box isn't able to properly support the weight of a ceiling fan. You must install an approved box before mounting the fan.

One approved box is simply a modified 4-inch by 1½-inch octagon box which has two holes drilled through its back, each directly above a cover-mounting screw hole. Long 8-32 screws are passed through the holes and are threaded as far as they will go through the tapped holes of the box, leaving two studs to which the fan mounting bracket will be attached.

The box is mounted on the face of a 2 by 4 block that's nailed between two ceiling joists, at a height that puts the front of the box flush with the finished ceiling. Four screws hold the box to the block. The mounting bracket of the fan is slipped onto the

studs and is secured in place with stop nuts. The fan is then mounted on its bracket according to the manufacturer's instructions.

There are kits available for mounting an approved ceiling-fan box in existing homes where there isn't access above the ceiling. Installation involves cutting a 4-inch round hole in the ceiling, and then passing an adjustable-length hanger through the hole and driving both ends of it into the ceiling joists. The box is then attached to the hanger.

Control for the fan can be a simple on-off switch or a speed control. An additional switch should be used when a light kit is installed on the fan. Some more expensive fans use a smart control that regulates the speed and direction of the fan as well as the lights, all from one wall-switch location.

Smoke Detectors

Every home must have at least one smoke alarm installed. A mix of 120-volt and 9-volt battery-operated units is best

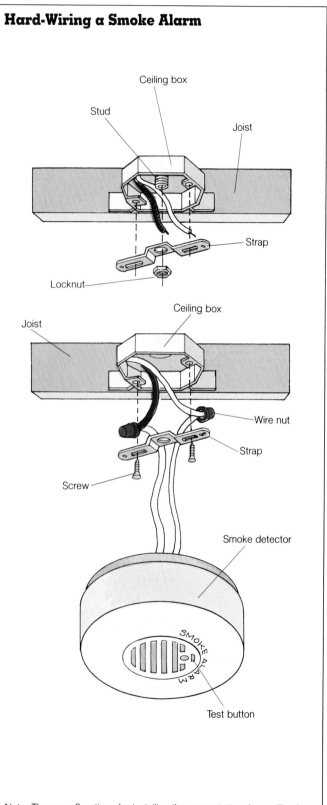

Hard-Wiring a Smoke Alarm

Ceiling box
Stud
Joist
Strap
Locknut
Joist
Ceiling box
Wire nut
Strap
Screw
Smoke detector
SMOKE ALARM
Test button

Note: There are 2 options for installing the support strap in a ceiling box: using a locknut or using wire nuts. Choose 1 option.

Installing a Doorbell

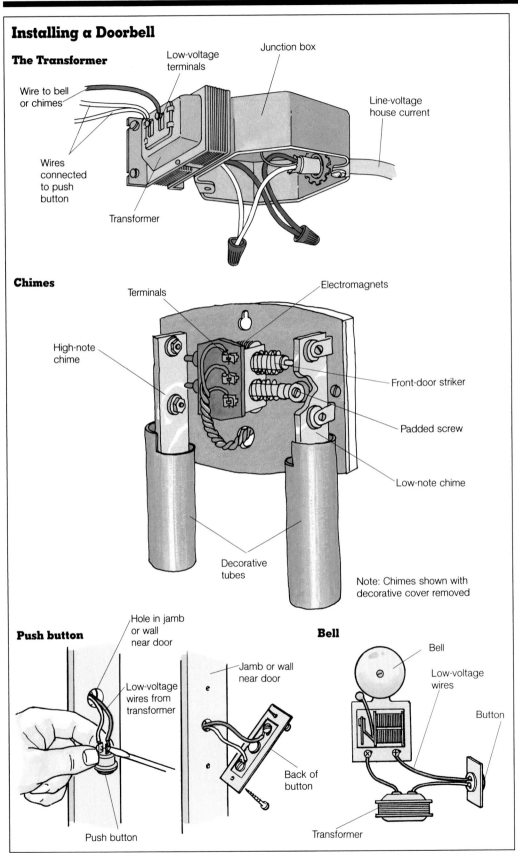

The Transformer

- Wire to bell or chimes
- Low-voltage terminals
- Junction box
- Line-voltage house current
- Wires connected to push button
- Transformer

Chimes

- Terminals
- Electromagnets
- High-note chime
- Front-door striker
- Padded screw
- Low-note chime
- Decorative tubes

Note: Chimes shown with decorative cover removed

Push button

- Hole in jamb or wall near door
- Low-voltage wires from transformer
- Jamb or wall near door
- Back of button
- Push button

Bell

- Bell
- Low-voltage wires
- Button
- Transformer

---

because there's little chance that the battery and the electric power will fail at the same time. Some manufacturers make 120-volt units that are backed up with a 9-volt battery. Smoke detectors should be installed near the furnace, in or near the kitchen, and in hallways outside of bedrooms.

Two basic sensing systems are used in detectors. Photoelectric systems are especially responsive to smoldering fires; smoke detectors of this type are best used in kitchen and bedroom areas. Dual ionization systems respond effectively to rapidly developing fires. This type is effective in furnace rooms and storage areas.

Motion-Sensor Lights

These special lights are generally used for outdoor installations. Whenever motion is detected within their range, the light automatically switches on. It stays lit for a set period of time and, if there is no further activity, turns itself off. Some systems can also discriminate by size, activated by a large dog, for example, but not a squirrel.

Motion-sensor lights are no more difficult to install than an ordinary light fixture. In fact, they generally follow the same procedures. Some are as straightforward as changing a light bulb.

Wiring a Three-Wire Appliance Receptacle

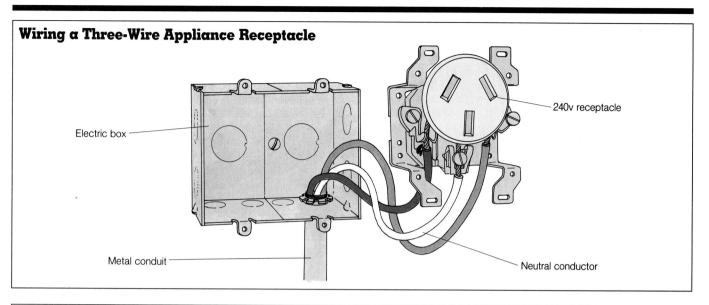

Electric box

Metal conduit

240v receptacle

Neutral conductor

Connecting a New Wire to an Existing Circuit

Note: Some areas may require pigtails. Check your local codes.

End-of-the-run receptacle

Middle-of-the-run receptacle

Middle-of-the-run switch

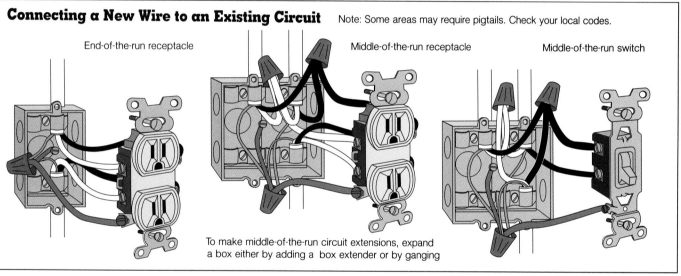

To make middle-of-the-run circuit extensions, expand a box either by adding a box extender or by ganging

Wiring a Multiwire Circuit

From source

Second receptacle is wired with red wires

Third receptacle is wired with black wires

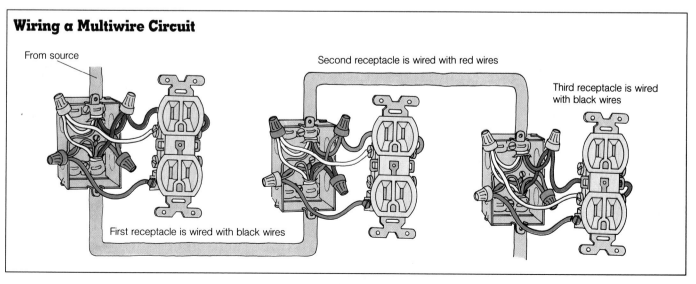

First receptacle is wired with black wires

OUTDOOR FINISH WIRING

Electrical connections in outdoor wiring are made in the same way as when wiring indoors. However, some areas of outdoor finish wiring require special attention.

Installing Fixtures

Although boxes for outdoor installations must be of weatherproof design, except those installed in exterior building walls, standard switches and receptacles can be used for the finish wiring and are installed using the same techniques as indoor installations. The covers, however, are different. Covers for duplex receptacles have either a pair of spring-loaded gasketed doors or a single door that covers both receptacles. GFCI covers have a single door. There are two options for covering switches. One cover has a single door that covers an ordinary toggle switch handle. The other type has no door. It has an auxiliary handle that is accessible at all times, but is electrically insulated from the switch and toggle that it conceals and protects. Install it by slipping a fork-shaped device, included with the cover, over the switch toggle and connect it to a shaft inside the cover. Then screw the cover in place.

Light fixtures are essentially the same as indoor lights, although they have gaskets or other weatherproof devices to ensure a tight seal against the box. Some have internal junction boxes built into them and are attached directly to conduit, without the need for a separate box.

Additional Weatherproofing

Great effort is made to make outdoor outlet boxes weathertight. Even so, in spite of gaskets and sealing methods, water from driven rain sometimes finds its way into these enclosures. Because this might happen, it's important to provide a way for the water to drain easily so it doesn't accumulate and cause damage. The simplest, most effective way to do this is to drill a ¼-inch weep hole in the bottom of the enclosure. This guarantees good drainage.

Box-sealing kits are available for boxes used outdoors. If you choose not to use such a kit when finishing, use a small amount of ordinary automotive grease on all threaded mechanical connections, and apply a light coat to all gasket surfaces. This makes it much easier to loosen device screws and box plugs in the future. Smear a light coat of petroleum jelly on all lamp threads before installing them. Do this and you should never have a problem removing a burned-out bulb.

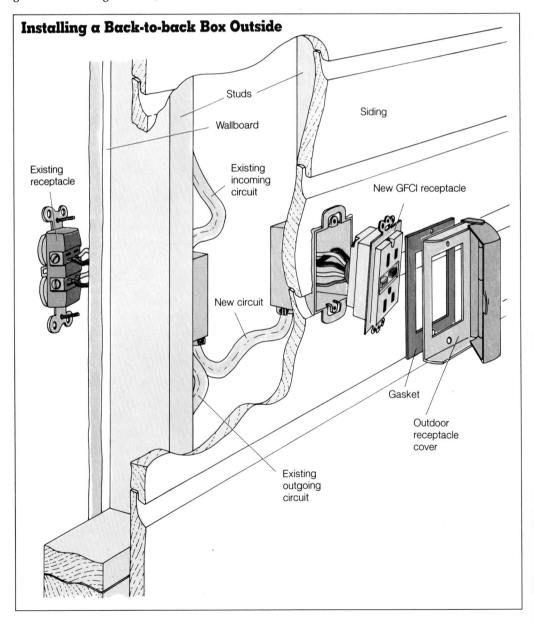

Installing a Back-to-back Box Outside

Studs

Siding

Wallboard

Existing incoming circuit

New GFCI receptacle

Existing receptacle

New circuit

Gasket

Outdoor receptacle cover

Existing outgoing circuit

Troubleshooting and Common Problems

Two simple tools can help you locate problems in electrical systems and equipment—the continuity tester and the neon test lamp. Either can be purchased for just a few dollars. Without these tools, troubleshooting can become little more than a guessing game.

A very common continuity tester has a cylindrical body that holds two small batteries and a small lamp. A 1-inch-long probe is attached to one end of the body, and a 3-foot-long test lead with an alligator clip on the outer end is attached to the other end. The light turns on when the alligator clip makes contact with the probe because this contact closes the circuit of the tester.

To check a questionable switch, first turn off the power. Then clip the test lead to one of the switch terminals, hold the probe to the other, and turn on the switch. If the switch is good, it will complete the tester circuit and the lamp will glow. A word of caution: Don't use a continuity tester on energized circuits. Besides the risk of shock to you, there's a risk to the tester. If the switch in the example above was in an energized circuit—a circuit complete with a properly operating light fixture—and the tester was applied to the switch when it was in the *off* position, 120 volts would be applied to the tester, and it would be damaged beyond repair.

Continuity testers can test the continuity only of circuits and devices that offer low resistance to the flow of electricity. This isn't a serious limitation because many electrical problems involve only switches and wiring interconnections. Successful use of the tester simply requires that it be applied systematically from point to point until the "open" is located. Testing other devices such as motor windings, solenoid valve coils, and heating elements is done with an ohmmeter, or the ohmmeter function of a multimeter.

When using a neon test lamp to check for voltage at a 120-volt receptacle, place one test lead in each of the parallel slots of the receptacle, and the lamp will glow if voltage is present. Remove the test lead from the longer of the two slots, touch the lead, and the lamp will again light. The lamp will also glow if you insert the open lead into the U-shaped slot of the receptacle, or if it contacts the metal screw that holds the wall plate in place. In fact, in each case where the test lead had been withdrawn from the larger slot, the slot to which the branch-circuit neutral is connected, the lamp would have glowed even if the neutral wire wasn't present at all. This illustrates the importance of testing at slots or terminals that connect to the current-carrying conductors of the branch circuit whenever possible.

Neon test lamps have many additional uses. Proper polarity at a 120-volt receptacle can be checked by making certain that voltage to ground is present only at the smaller of the two parallel slots. To test this, hold one of the leads while you insert the other first into one slot, then into the other. If the polarity is all right, place one lead back into the small slot and the other into the U-shaped slot; if the neon glows, the grounding at the receptacle is okay, too. Most test lamps will operate at voltages from 60 to 250 volts, so they can also be used to check your 240-volt circuits.

Replacing a Ballast

Installing a new ballast in a fluorescent fixture is a common project. First make certain the ballast is the problem by checking the circuit for power, changing the lamps, and the starter if the fixture uses one. If this doesn't restore operation, disconnect the power to the fixture and immediately open it up to expose the ballast. Carefully touch the ballast to see if it is warm. If it isn't, it probably needs to be replaced.

Begin by removing the existing wiring to the lamp holder and power source at a point close to the ballast.

Remove the defective ballast—usually only two screws hold it in place—and install the new unit. Tighten the mounting screws or nuts, so that the ballast is held very firmly to the fixture. Be certain that the black and white wires on the ballast are toward the same end of the fixture as those of the original ballast were.

The wires on replacement ballasts are seldom long enough to reach the lamp holders, so it's common practice to splice the old and new wiring with wire nuts. It's not always possible to connect the wires to the lamp holders color for color because over the years color standards have changed. Also, any high temperature wire must be replaced with the same kind. To avoid problems, follow the wiring diagram shown on the ballast nameplate. Extend each of the new wires toward its destination, then fold the final 4 inches of each wire back toward the ballast. Extend the old wires toward the ballast and cut off each at a point next to the end of the new wire to which it will be connected. Strip ½ inch of insulation from the ends of the wires and connect old to new using wire nuts. Bundle the groups of wires on each side of the ballast and band each group together with two or three wraps of electrical tape.

Types of Plugs

Screw Terminal Plug

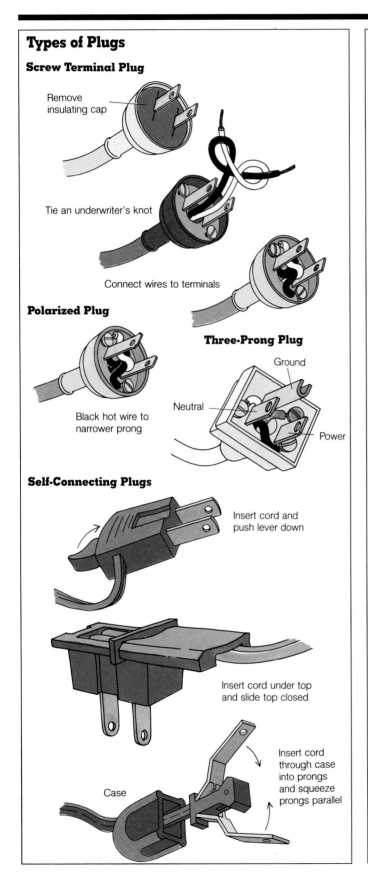

Remove insulating cap

Tie an underwriter's knot

Connect wires to terminals

Polarized Plug

Black hot wire to narrower prong

Three-Prong Plug

Ground

Neutral

Power

Self-Connecting Plugs

Insert cord and push lever down

Insert cord under top and slide top closed

Case

Insert cord through case into prongs and squeeze prongs parallel

Fluorescent Fixture Wiring

Rapid Start

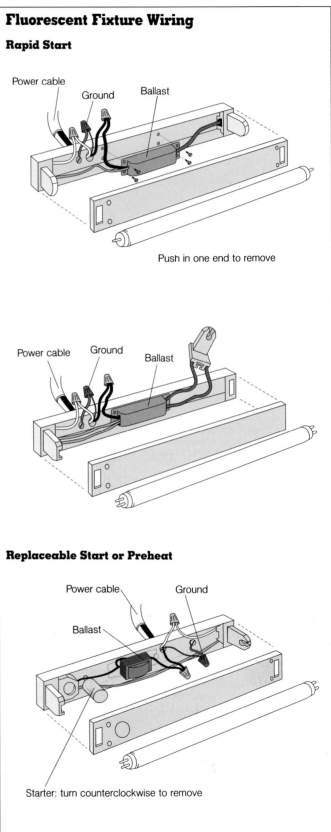

Power cable

Ground

Ballast

Push in one end to remove

Power cable   Ground   Ballast

Replaceable Start or Preheat

Power cable

Ground

Ballast

Starter: turn counterclockwise to remove

Lamp Components

Finial

Outer shell

Insulating sleeve

Socket

Socket cap

Underwriter's knot

Shade

Harp

Socket

Threaded tube

Base

In a lamp with 1 socket, the cord goes into the base, runs up through the threaded tube, and connects directly to the socket terminals.

In a lamp with 2 or more sockets, each socket is wired separately, and the wires are connected to the main cord with the wire nuts at the top of the threaded tube.

Rewiring an Incandescent Lamp

Lamps with incandescent bulbs are all quite similar, whether they are large floor lamps or small table models. A lamp has only a few basic parts that can cause a malfunction: the bulb, the socket, the switch, the cord, or the plug. Understanding how a lamp is put together not only allows you to make any necessary repairs, but also to make your own lamps from old bottles, pieces of driftwood, or anything that strikes your fancy. All the parts to put a lamp together from scratch can be found in kit form in hardware and electrical stores.

Taking Apart a Lamp

1. Unplug the lamp and remove the bulb.

2. Remove the shade. It may be necessary to unscrew the finial at the top of the harp to free the shade. If your lamp has a 2-piece harp, slide the 2 metal sleeves up at the base of the harp, squeeze the harp at the base, and lift it out.

3. The socket consists of 4 pieces: the outer shell, the insulating sleeve, the socket with switch and terminal screws, and the socket cap. To remove the shell, press in with your thumb where the shell is stamped "press" and lift up. If the shell will not come off, pry it up with a screwdriver. Remove the cardboard insulating sleeve. Loosen the 2 terminal screws on the socket and remove the wires. Lift the socket out of the cap.

4. To remove the cap, first loosen the setscrew at the base of the cap. Unscrew the cap from the threaded nipple. If there is no setscrew and the cap does not readily unscrew, give it a forceful turn.

5. If you must remove the cord from the lamp pipe, tape the new cord to the old and pull it through as you remove the old cord.

6. Part the new cord about 2″ back and tie an Underwriter's knot at the top, as illustrated.

7. Strip off ½″ to ¾″ of insulation at the end of the wire, twist the wire clockwise, and wrap it clockwise around the terminal screws. Tighten the screws. Reassemble the lamp.

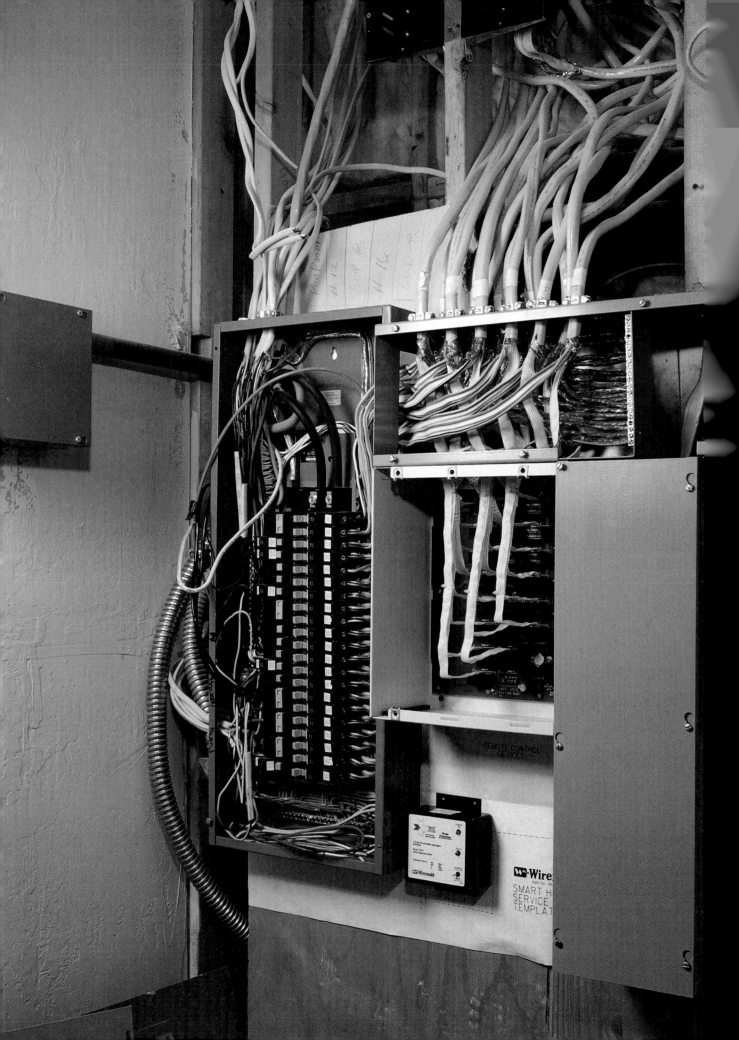

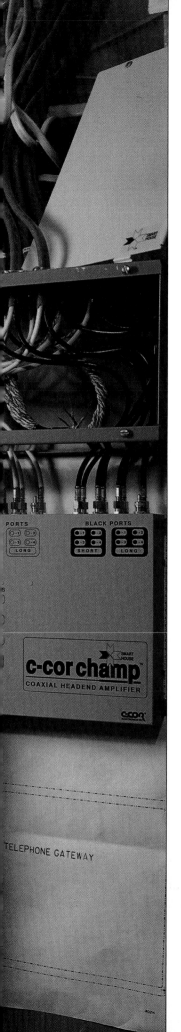

ALTERNATIVE SYSTEMS AND SMART HOUSES

Electricity is a uniquely flexible form of energy, and uses for electric power continue to expand. This expansion creates a demand for power companies to increase production and add facilities. At the same time economic, political, and environmental pressures are forcing power companies to restrict expansion or explore alternative ways to produce electricity. For example, many companies already have large solar or photovoltaic systems feeding their AC distribution grid.

On the residential level, energy-saving appliances and lighting systems help reduce overall demand. Many homes now have complete low-voltage systems, either as a backup or as the sole electrical system. Another emerging development is Smart House technology, which combines standard AC wiring, low-voltage wiring, electronic components, and computer-based controls into one integrated system. Although not all of these developments may gain widespread use, some of the systems presented in this chapter may help you conserve energy, solve unique problems, or assess future options in your long-range planning.

This Smart House load center includes much more than the standard circuit breaker panel on the left. The yellow hybrid cables carry both 120v AC current and low-voltage control messages. The orange cables distribute control communications and 12v DC current for switches, smoke detectors, security systems, and other low-voltage applications. The coaxial network links video, audio, and computer equipment and shares communications cables with a telephone gateway to be installed.

CONSERVATION

Many utilities have rebate programs for home insulation, water–heater blankets, hot–water–pipe insulation, energy efficient appliances, and replacement of incandescent lamps with compact fluorescents. Utility companies are adopting these techniques to reduce costs, loads, pollution, and even the building of new generating plants.

Lighting

Incandescent lighting is estimated to consume 25 percent of all the electricity generated in the United States. Ordinary fluorescent lamps produce an estimated 20 percent of the heat load in an office building. Compact fluorescent lamps, on the other hand, manipulate the AC power with electronic controls to yield more light while consuming up to 70 percent less power. Compact fluorescents also produce better light, almost no heat, very little radiation, and can have a life span 10 times longer than incandescent lamps.

Because one 75-watt incandescent lamp will use a barrel of oil at the generator during its lifetime, with that barrel of oil producing a ton of carbon dioxide during combustion, using compact fluorescents can reduce generating costs and pollution. These savings result not just from lighting demands but because of the greatly reduced cooling demands as well.

In 1991, California announced a program for state buildings to replace over three million incandescent light bulbs, costing 75 cents each, with compact fluorescents costing about $18 apiece. The initial expense, including automatic motion-controlled light switches and other amenities, was $90 million, but the calculated savings in reduced energy use is $765 million over the next five years. The reduction in pollution associated with the reduced consumption is "equivalent to removing 200,000 cars from the state's highways," according to then-governor Deukmejian.

Other U.S. studies indicate that if all the incandescent bulbs were replaced with compact fluorescent lamps tomorrow, 18 nuclear plants could be eliminated the day after.

Heating

Electric "baseboard" heaters are very popular for retrofit installations and additions because they are clean, easy to install, and eliminate the duct work necessary to connect with a central furnace. Most of these heaters are very inefficient and slow to respond, however. New types of electric radiant-panel heaters, long used in Europe, are more efficient, just as clean, and even easier to install; available for installation in ceilings and floors, the improved performance over baseboard types is dramatic, and the efficiency documented. Only unfamiliarity with these devices limits their use in the United States, and this is changing.

Appliances

Refrigerators consume about 20 percent of estimated electricity use in the United States. New refrigerators carry a sticker that indicates average annual operating costs. The real costs vary, depending on how the appliance is used and the utility rates in the area, but this information does permit a relative comparison of available choices. Utility rebates are often based on this relative efficiency; you can buy a better refrigerator, get a larger rebate, and also pay lower operating costs over the life span of the appliance.

Electric water heaters also consume a lot of power and bear relative-consumption stickers. Rebates are available in some areas, but common-sense improvements can also reduce consumption. An insulating blanket around the heater and hot-water-pipe insulation can reduce costs dramatically as well as improve the functioning of the system. These blankets and pipe insulations are an inexpensive, one-time purchase. Paying for heat loss suffered without these items continues month after month, year after year. All wasted heat consumes money that needn't be spent.

Although the relative-cost figures on new appliances are helpful, they are based on average rates. Not only will these rates change, they are not comparable in all geographic areas. In the Northwest United States rates are low compared to the Northeast because of the tremendous hydrogenerating capacity and relatively small population in the Northwest. Readers in the high-rate areas have an added incentive to buy the most efficient appliances they can afford and to consider renewable source systems.

Energy-Saving Light Bulbs

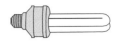

Compact fluorescent

Fluorescent diffuser lamp

Fluorescent spot lamp

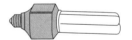

Panasonic twin-tube capsule

Electric light capsule

BACKUP SYSTEMS

Twelve-volt electrical systems are gaining in popularity. Improvements in storage technologies developed for 12-volt renewable systems where there is no AC grid, and systems designed to protect computer data, have made this technology useful for backup systems even in developed areas.

Residential Systems

An emergency backup system for a home is relatively inexpensive insurance in case of disaster, and can be extremely reassuring when the power grid goes out. The system, consisting of a transformer, charge controller, and deep-cycle battery, can be located in the garage, basement, or some other vented location. The only maintenance this setup requires is filling the batteries three or four times a year. It stands ready to supply a 12-volt radio, TV, auxiliary lighting, and a citizens band radio. The whole system currently represents an investment of less than $500, including the radios and a small black-and-white television.

These systems are recommended for homes located in hurricane country, class 4 seismic zones, or other areas prone to natural disasters. There are numerous examples of such backup systems performing flawlessly during the aftermath of the 1989 Loma Prieta earthquake in California. They can be very reassuring to homeowners until the grid power and communications are resumed.

Computer Backup Power Supplies

Hospitals, police, and fire stations have long relied on expensive automatic generator backup systems in case of grid failure. Generator-fed systems are very noisy and costly to operate and maintain, but were necessary to protect the extremely valuable computer data of these organizations. Now another option has emerged.

A UPS, or uninterruptable power supply, is a battery-based system that automatically produces AC backup power for a computer upon AC grid failure. Many large manufacturers are installing massive UPS systems for protection of computer controls of assembly-line robots, and smaller systems for delicate control components.

Small, packaged UPS systems are also available, consisting of a 12-volt (gel) battery, an inverter to change the 12-volt DC into 120-volt AC, and an off-grid warning light or beeper. The current $500 models will provide approximately a half hour of backup power, enough time to organize and save the work in progress. Readers considering home backup systems can also protect their computers by adding an inverter and warning device to their emergency systems at little additional cost.

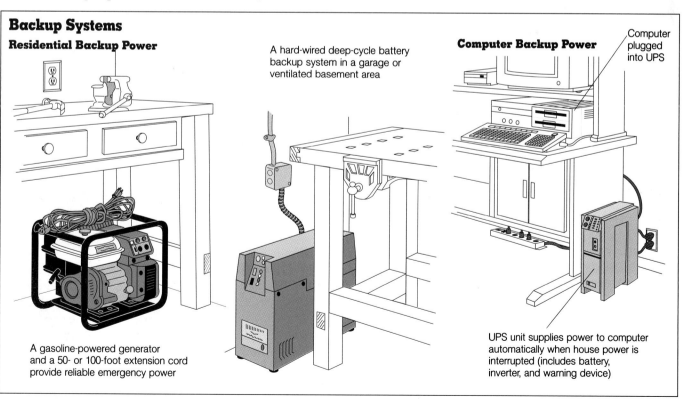

Backup Systems

Residential Backup Power

A hard-wired deep-cycle battery backup system in a garage or ventilated basement area

A gasoline-powered generator and a 50- or 100-foot extension cord provide reliable emergency power

Computer Backup Power

Computer plugged into UPS

UPS unit supplies power to computer automatically when house power is interrupted (includes battery, inverter, and warning device)

RENEWABLE SOURCE SYSTEMS

Stand-alone systems are becoming economically attractive alternatives wherever there is enough sunlight, water, or wind to supply such a system. With current costs, a stand-alone photovoltaic system will begin payback within 8 to 10 years, even sooner if existing AC wiring can be used or you do the rewiring work yourself.

General Considerations

Stand-alone itself is a misnomer. Renewable and AC grid systems may be combined, and some states even have laws that require the local grid utility to pay for any electricity fed from the renewable system to the utility grid. In these states, money spent on more generating capacity and suitable storage is returned after the break-even point, when the utility will send regular payments to the system owner for the excess generated power.

Managing these systems is very simple and takes on average less than one hour every couple of months. The only difficult part, changing defective batteries, should occur only once every five to seven years.

System configuration will vary depending on needed capacity, whether 12-volt or 120-volt appliances will be used, and if surplus power can be sold to the utility. Readers interested in renewable systems are referred to the growing number of books on the subject, but a brief description here of components and how they are used reveals a lot of possibilities.

Photovoltaic Panels

Thousands of small cruising vessels as well as the Coast Guard use photovoltaic-supplied navigational aids, and the U.S. Geological Survey uses photovoltaic-supplied instruments for earthquake-monitoring devices. The reliability of these photovoltaic systems is impressive.

Photovoltaic technology uses modular panels made of silicon-based materials that turn both direct and indirect sunlight into DC electrical flow. Although photovoltaic panels suffer from both transfer and leak losses, direct sunlight delivers a staggering 100 watts per square foot when it strikes the earth. The group of panels, known as an array, may be mounted on the roof of a house or located away from the dwelling. Panel costs have decreased greatly during the last 10 years, and new materials are likely to continue this trend.

Control Features

Because direct sunlight generates more electricity in a panel than sunlight striking at an angle, passive tracking devices are available. These devices consist of a support post, a rack to mount the panels, and two small pressurized gas bottles.

These bottles are installed on the support post so that as the sun moves, one bottle will be exposed to the light. This bottle will heat up, driving the gas to the other bottle. The pressure change moves a pneumatic cylinder that adjusts the position of the tracker. The whole operation is automatic and reliable.

The DC electricity from the panels is fed to storage batteries until the power is needed. Linear modifiers adjust panel output current to suit the storage portion of the system.

Charge controllers are simple devices that limit the incoming DC supply from the array. More elaborate controllers also provide a central point to gauge the capacity and condition of the system. The fancy controllers are expensive and currently used mainly in situations where photovoltaics are the only power source.

Inverters are special transformers that electronically convert 12-volt DC electricity into standard 120-volt 60-hertz AC current with minimal power loss. This allows a bank of 12-volt batteries to power standard 120-volt appliances and normal house wiring. Inverters are expensive but very reliable, and should be considered a one-time purchase. Inverters without fans are recommended because of the decreased noise. Some inverters do have a slight but constant power loss, and they are turned off automatically, using a time switch, or by hand at the end of the day.

Storage

To provide electricity suitable for home use, more durable and long-term storage banks are required. Units known as deep-cycle batteries are used,

the storage capacity of which has more to do with system performance than other factors. Typically, storage requirements are calculated to provide electricity after three full days without direct sunlight. Where a 180-amp-hour marine deep-cycle battery (about $80) is perfect for a backup system, stand-alone systems will usually benefit from using more exotic and expensive batteries, such as the so-called chloride types which currently cost more than $1,000 for a 400-amp-hour bundle of six cells. Other kinds of batteries fall in between these extremes in both price and performance, and new and better methods loom on the horizon.

12-Volt DC Systems

The primary advantage to using inverted AC power from the storage bank is that existing AC wiring may be used to supply the existing appliances. But in situations without an established AC wiring system, direct use of DC output should be considered. The drawback to a 12-volt-only system is that delivery of sufficient wattage requires much more amperage and therefore much larger wires than an AC system. DC wiring is also more expensive than AC materials and harder to install, but no inverter is required. Electric razors, soldering irons, blankets, hair dryers, washing machines, and even refrigerators and heaters that use 12-volt DC power are currently available. Twelve-volt lamps, fans, radios, tape decks, and so forth, can be purchased from auto-supply stores, and water pumps are available from marine suppliers.

DC Distribution Panels

Due to the increased amperage found in a straight 12-volt system, different panels and breakers are used. Marine versions are suitable for small-load applications only, such as a backup system. Larger systems require using a panel and breakers designed for the purpose. The breakers are inexpensive but DC panels are rare and sell for about 10 times the cost of a similar AC panel. These DC panels need bigger lugs to tie larger wires to the buses and are manufactured in 12- and 24-volt configurations.

Small Hydro Systems

This technology is less expensive than photovoltaic power generation, but a constant source of water of suitable quality must be available to develop such a system. A small stream is seldom sufficient as a source, and tapping one will not benefit the stream or the fish in it. The critical source qualities necessary are volume and head. Head refers to the vertical drop between the source and the turbine that generates the power. The greater the head, the less water needed to produce pressure at the turbine. With 400 feet of head, small nozzles will consume 4 gallons a minute to produce 100 watts of 12-volt power; with the minimum 12-foot head, larger nozzles must be used to generate 100 watts, and 400 gallons per minute will be consumed. A hydro turbine will operate 24 hours a day. The turbines currently retail for under $1,000.

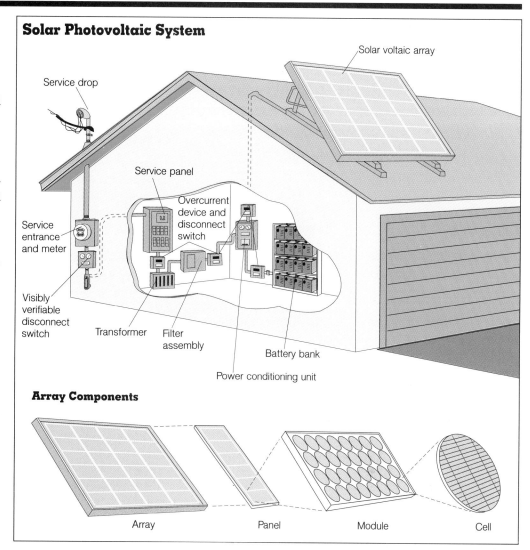

Solar Photovoltaic System

Service drop

Solar voltaic array

Service panel

Overcurrent device and disconnect switch

Service entrance and meter

Visibly verifiable disconnect switch

Transformer

Filter assembly

Power conditioning unit

Battery bank

Array Components

Array

Panel

Module

Cell

Wind Generators

These represent the least-reliable alternative power source. The most efficient wind generators require an average wind speed of 5 miles per hour just to start generating power. Unless the usual wind is such that you can barely walk from the house to the car without getting blown over, wind systems will generally prove to be an unwise investment. They may be advantageous for intermittent use, such as for pumping water into a storage tank.

The *NEC*

The 1990 *NEC* requires separate overcurrent devices between the DC source and the charge controller, the battery bank, filters (linear modifiers), the inverter, and between the inverter and main AC panel. Although two-wire DC circuits are not required to use grounding circuit wires, roof-mounted photovoltaic arrays must have GFCI protection tied to the metal frame of the array. These rules apply to both hybrid and stand-alone systems; the protection afforded to the expensive components is worth the costs even in stand-alone applications. As long as the renewable-system voltage does not exceed 50 volts, this is the limit of the *NEC's* impact on these systems. As this technology gains in usage, however, it is likely to become more regulated.

One last caution: Traditionally, AC systems use the black wire to carry power to outlets and the white wire to return potential to zero. With low-voltage DC current, however, it is the red wires that carry the power to the outlet and the black wires that return the potential to zero.

SMART HOUSES

Innovative home-wiring technology that incorporates improved safety, greater efficiency, and control features and options that have not been possible before is emerging. This system is referred to as Smart House and is the product of a collaboration of more than 40 different groups, including utilities, software designers, and others.

Efficiency

The electronic miniaturization that makes personal computers relatively inexpensive today can be used to manage and control an electrical system as well as the heating, cooling, and other systems in a home. This control can reduce energy use and costs by running the large loads when demand (and cost per kilowatt-hour) is at its lowest. Overall costs are further reduced because the wiring for so-called optional systems, such as telecommunication, intercom, cable TV, and security systems, is included in the Smart House cables, eliminating subcontractors and the need to pull the optional wiring separately. These cables are altogether different from what is commonly used today and can perform many other functions as well. The boxes and panels and most devices are also different. Only the large dedicated-circuit loads will retain the Type NM cables and receptacles familiar to home electricians. The higher initial costs of this approach are offset by the reduced energy consumption and the dramatically improved convenience.

By using the control features to run large-load appliances in the middle of the night, the homeowner pays the lowest possible rate for the energy consumed. By using low-voltage feedback controls, a dishwasher can turn up the water heater just before the cycle begins and turn it down after the cycle is over. Together these controls can turn up the water heater a half hour before shower time on weekdays and back down at other times, saving even more energy.

Modern programmable thermostats are simply reactive devices. Smart House heating and cooling systems, on the other hand, provide not only their basic functions, but can measure inside and outside temperatures and check the date and time of day to anticipate changes in these conditions. Smart House controls will also close the drapes on hot summer days on windows with southern exposure and open them on sunny, cool winter days. Heating and cooling efficiency is improved further by heating only the most frequently used areas of the home. Miniature thermostats and low-voltage fans or baffles in the ducts can direct forced-air heat only where it is needed, bypassing other areas. This is called zonal heating and can reduce consumption by 30 percent.

Safety

A Smart system will link the 120-volt appliances in a home to special breakers in the panel so that until the appliance is turned on, no AC power will flow through the circuit. This reduces electromagnetic radiation (EMR) while protecting children from shocks from unprotected receptacles. Should a problem develop in an appliance, the Smart technology can notify the homeowner on site or with a phone call, and turn off the power to that specific appliance.

All the control features use low-voltage DC electricity, further reducing EMR fields. One cable also contains two larger DC conductors, and these are intended not only to supply power to electronic memories and so-called phantom loads (clocks, timers, security systems), but to serve DC appliances directly as well. Common home electronics, such as TV, radios, and tape decks, already transform the grid AC into DC power, so that the peaks and valleys of AC current don't show up as snow or noise in the picture or sound.

Navigating through your house at night is made remarkably safe and easy. A touch of the control panel will light a path through the house to your destination.

If you are home alone and the doorbell rings, there is no need to go to the door. You can view the visitor from anywhere in the house and issue instructions to him or her. The system can even be set up to call you at your office if there is no one home and the doorbell rings.

Entertainment and Communications

Smart system design provides for TV signals with a pair of special coax cables dedicated to audio and video signal transmission. This cable can also be used to direct a VCR signal to any TV in the home, and with a video camera, monitor a sleeping baby during commercials or on a split screen. Other wires in the same cable allow the speakers of an audio system to be placed in any room, and the sound to be controlled from anywhere in the home.

Smart cabling has provisions for intercoms and security built in. The telephone system includes a new control box, called a gateway, that has multiline capability and a built-in digital answering machine. Messages can be reviewed through the TV during commercials. In addition, a fiber-optic cable-ready link is provided in the box, anticipating future use of this technology. In a Smart House, the smoke detectors may be connected through the gateway to

Smart House

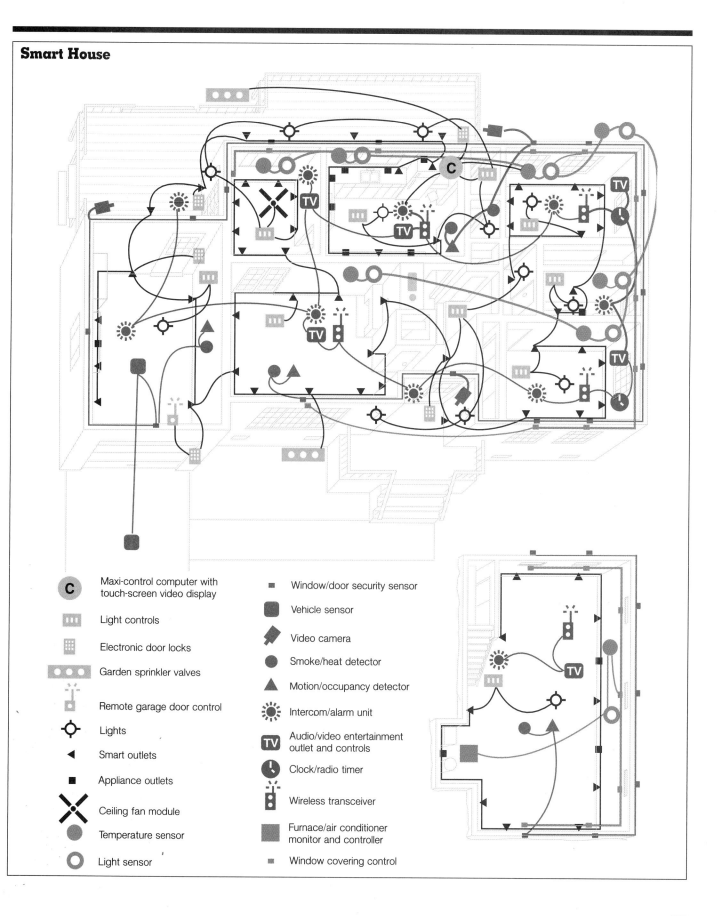

C	Maxi-control computer with touch-screen video display	■	Window/door security sensor
	Light controls	■	Vehicle sensor
	Electronic door locks	🔦	Video camera
	Garden sprinkler valves	●	Smoke/heat detector
	Remote garage door control	▲	Motion/occupancy detector
⌀	Lights	☀	Intercom/alarm unit
◀	Smart outlets	TV	Audio/video entertainment outlet and controls
■	Appliance outlets	🕐	Clock/radio timer
✕	Ceiling fan module		Wireless transceiver
●	Temperature sensor		Furnace/air conditioner monitor and controller
○	Light sensor	■	Window covering control

contact the fire department automatically via phone, informing it of the problem and the address.

A home automation system can also use the normal assortment of motion detectors, infrared-beam detectors, and sonic detectors already on the market to graphically pinpoint the location of an intruder on any video screen. With a mini-cam concealed in each room, the intruder can even be seen on screen.

Control Systems

All this flexibility and the new features would be impossible without the computer control that interconnects these different circuits in a logical way. The Smart House system is so refined that the user needs no computer knowledge at all. It's simply a matter of viewing a list of options on a TV screen and picking the desired function by touching the screen. Voice-activated controls are being developed for the sight impaired and for the physically challenged.

Another control feature allows the user to key any telephone on site and tell the system what to do. TV and VCR remote controls may also be used to operate the whole system; and small touch-tone keypads can be placed where needed for system control. Unlike the touch-screen control, users need to remember the

relationship between commands and various functions, but if the telephone control is used, it will function from a telephone anywhere in the world.

These control functions are just a few of the possibilities. Others include sprinkler controls, skylights that open and close automatically, and virtually every household function that can be automated. The significant difference with Smart control features is that they may be easily programmed to perform these and other complex functions automatically, with different cycles for weekday and weekend schedules, and the programmed functions can be added to or modified using quick, simple methods.

Backup Power

Smart control features are provided with an electric backup so that once the basic functions of the system are established, these functional commands are held in the system memory even in the event of loss of grid power. This is identical to the small UPS system described earlier, except the power is used to protect the memory of the whole system and not just a single computer.

Brownout Protection

In areas where brownouts occur at times of peak demand, some states have laws that allow the power company to disengage large appliances such as dryers and air conditioners from the grid for short periods. These voluntary programs require special devices to accomplish this, a job that Smart controls will do better. Besides the reduced costs incurred, power companies usually provide an annual rebate to users as well. After a peak overload, there are lots of appliances set to start again simultaneously, further straining the generating and transmission systems of the grid. With the link between the utility and the user provided by the Smart House interface, utilities can phase in the restored power gradually and eliminate this shock to their systems.

Convenience

Although home automation will cost more initially, the benefits of improved safety and efficiency, and the tremendous flexibility built into the system, are not just another step forward, but an entirely different approach to home wiring. The homeowner will be able to use and control the formerly independent operations as a single system, accomplishing complex tasks easily.

By integrating the home systems (and renewable on-site system) with the utility grid, a daily check on household power consumption can be displayed on the TV each evening, and the Smart system can be instructed to continue, increase, or decrease consumption the next day.

Hybrid Systems

Although the Smart House is a packaged product, these options can be installed piecemeal, according to site-specific opportunities and use requirements. In other words, although most homeowners must depend on utility-supplied AC current, elements from photovoltaic, small hydro, and wind-driven systems can be grafted into an AC residential system fairly easily. As the cost of utility-generated power increases, reliability decreases, and pollution gets worse, hybrid and independent systems look better and better.

Proposed uses of new systems appear wide-ranging and thorough, and other uses and options will undoubtedly be incorporated as smart technology continues to evolve. During its short life, the ways we use electricity have changed dramatically and some mistakes have been made in the process. This pattern is not likely to change; remember this is an evolving technology. Even with the advent of Smart systems, there is plenty of room in the technology for creative thought and innovative solutions.

GLOSSARY

Alternating current (AC) An electric current in which voltage flows in one direction one instant and in the opposite direction the next. The direction of current flow reverses at regularly recurring intervals.

Ampacity How much current, expressed in amperes, a conductor can carry.

Ampere A measure of the amount of electric current flow.

Applied codes The most recently adopted *NEC*, plus locally required additions to the *NEC*, at the time of the work. Also known as Enforced Codes and Local Codes.

Approved Minimum standards established by an authority have been met.

Armored cable A flexible, metallic-sheathed cable used for indoor wiring, commonly called BX or Greenfield.

Ballast A magnetic coil that provides the starting voltage or stabilizes the current in a circuit (as of a fluorescent lamp).

Bonding The permanent adhesion of metallic parts, forming an electrically conductive path. *See also* Ground.

Box A device used to contain wire terminations where they connect to other wires, switches, or receptacles.

Branch circuit A circuit that supplies a number of receptacles for lights or appliances.

Burrs The rough edges caused when cutting conduit pipe.

Bus bar A rigid conductor at the main power source to which three or more circuits are connected.

BX *See* Armored cable.

Cable A conductor consisting of two or more wires that are grouped together in an overall covering.

Chases Spaces inside finished walls and between floors used for running ductwork or vent pipes.

Circuit The complete path of an electric current, leading from a source (generator or battery) through components (for example, electric lights), and back to the source.

Circuit breaker An electromagnetic or thermal device for interrupting an electric circuit when the current exceeds a predetermined amount; can be reset.

Code, National Electrical (NEC) A set of rules sponsored by the National Fire Protection Association, under the auspices of the American National Standards Institute, to protect persons, buildings, and their contents from dangers due to the use of electricity.

Color coding The identification of conductors by color.

Common ground connection Where two or more continuous grounded wires terminate.

Conductor A low-resistance material, such as copper wire, through which electricity flows easily.

Conduit Metal or plastic tubing used to enclose electrical conductors.

Connector, solderless A device (typically plastic insulated) that uses mechanical pressure rather than solder to establish a connection between two or more conductors; wire nut.

Continuity An uninterrupted electrical path.

Current The movement or flow of electricity; the time rate of electron flow, usually measured in amperes.

Cycle A complete positive and negative alternation of a current or voltage.

Device A unit or component that carries but doesn't use current, such as a junction box or switch.

Direct current (DC) An electric current that is not pulsed and flows in only one direction.

Doubletaps Situations in which one fuse or breaker is used to protect two circuits.

Electric charge The electric energy of a body or particle. The electron has an inherent negative charge; the proton has an inherent positive charge.

Electric current The flow of electricity through a conductor.

Electron The negatively charged particle of an atom; also called negatron. The flow of electrons in a conductor is what constitutes electric current.

Exposed Wiring designed to be easily accessible.

Feeder The circuit conductors between the service equipment and the branch-circuit overcurrent device.

Fish tape Flat, steel spring wire with hooked ends, used to pull wires through conduits or walls.

Fitting An accessory (such as a bushing or locknut) used on wiring systems to perform a mechanical rather than electrical function.

Flexible cable A conductor made of several strands of small-diameter wire.

Four-way switch Used along with two 3-way switches to control one or more lights from three separate locations.

Fuse A safety device inserted in series with a conductor, containing metal that will melt when heat is produced by an excess current in the circuit, thereby breaking the circuit.

Ground A conducting connection between an electric circuit and the earth or some other conducting body.

Grounded Connected to the earth or some other conducting body. The grounded wire is always white.

Grounding conductor The wire (green or bare) in a cable that carries no current; used as a safety measure to establish a ground.

Ground fault A situation where electricity can flow outside the conductors intended to carry power: When a hot wire at a bare point, for example where connected to a receptacle, touches a grounded component, such as a conduit or a grounding wire.

Ground Fault Circuit Interrupter (GFCI) Sophisticated overcurrent device that detects minute leaks of power and then quickly disengages that circuit.

Hanger A metal or insulated strap used at intervals to support electrical cable between points of connection.

Hot wires The conductors of a house circuit that are not grounded and are carrying power; any color but white or green.

Impedance The total opposition to alternating current by an electric circuit.

Insulation Nonconducting materials used to cover wires and in the construction of electrical devices.

Insulator A nonconductor that is used to support and isolate a conductor that carries current.

Inverter Electronic transformer capable of converting DC electricity into AC power at high levels of efficiency.

Junction box A box in which several conductors are joined together.

Kilowatt-hour (Kwh) The standard billing unit, equivalent to the energy transferred or expended in one hour by one kilowatt of power. Equals 1,000 watt-hours.

Knockout A die-cut impression in outlet and switch boxes so designed that it can readily be removed to provide an opening for access.

Lighting outlet An outlet allowing the direct connection of a lamp holder, lighting fixture, or pendant cord ending in a lamp holder.

Live wire   A wire that carries current; also called hot wire.

Lugs   Clamping elements built in to electrical devices.

Meter, electric   A device that measures how much electricity is used.

National Electrical Code   *See* Code, National Electrical.

Neutral wire   The conductor in a cable that is kept at zero voltage. All current that flows through the hot wire must also flow through the neutral wire.

Ohm   The unit of resistance in a circuit or electrical device; equal to the resistance in a conductor in which one volt of potential difference produces a current of one ampere.

Open circuit   An electric circuit with a physical break in the path (caused by opening a switch, disconnecting a wire, burning out a fuse, and so forth), through which no current can flow.

Outlet   A metal or plastic box in which electrical wiring is connected to electrical components.

Overcurrent-protection device   A fuse or circuit breaker that is used to prevent an excessive flow of current.

Overload   Current demand exceeding that for which the circuit or equipment was designed.

Plug, attachment   A device inserted in a receptacle to connect conductors of attached flexible cord and conductors that are permanently connected to the receptacle.

Polarity   See Polarizing.

Polarized plug   A plug whose prongs are designed to enter a receptacle in only one orientation.

Polarizing   Using color to identify wires throughout a system to ensure that hot wires will be connected only to hot wires and that neutral wires will run back to the ground terminals in continuous circuits.

Power   The rate at which work is being done.

Raceway   A channel that contains electrical conductors.

Receptacle   A contact device installed at the outlet to supply current to a single appliance.

Resistance   A property of an electrical conductor, measured in ohms, that resists the flow of current according to the atomic nature of the conductor. Conductors such as copper, silver, and aluminum offer little resistance; poor conductors, such as glass, wood, and paper, offer more resistance.

Romex®   A trademark for one brand of NM cable (nonmetallic-sheathed cable) used for indoor wiring.

SLB (Service LB)   El-shaped conduit body with an opening on the back for wiring access, designed to mount flush to a wall.

Screw terminal   A means for connecting wiring to devices using a threaded screw.

Service conductors   Electrical conductors that extend from the street main or transformer to the service equipment of the building being supplied with power.

Service equipment   Electrical equipment located near the entrance of supply conductors that provides main control and enables cutoff (via fuses or circuit breaker) for the supply of current to the building.

Service panel   The main panel through which electricity is brought from an outside source into a building and distributed to the branch circuits.

Short circuit   An improper connection between hot wires or between a hot wire and a neutral wire.

Smart house   An emerging technology in which computers are used to incorporate telecommunications, entertainment consoles, energy efficiency, and safety features into the electrical system of a home.

Solar photovoltaic systems   Control and storage devices that convert sunlight into electricity.

Solar panels   Slices or thin films of semiconductor material that are wired, assembled, and bonded to a suitable substrate to form a single panel of uniform size and optimum output.

Splice   A connection made by joining two or more wires.

Split receptacle   A dual receptacle in which each of the two outlets is connected to a different branch circuit rather than a common circuit.

Stranded wire   A quantity of small cables that are twisted together to form a single conductor.

Switch   A device used to connect and disconnect the flow of current or to divert current from one circuit to another; used only with hot wires, never with neutral or ground wires.

Symbols, electrical   Lines, letters, and signs used on building plans to represent where wiring circuits, switches, receptacles, and other electrical features are located.

Three-way switch   A switch having three different conductor terminals. Two such switches are needed to control a light from two separate locations.

Triplex   Cabled assembly of two insulated conductors loosely wrapped around one bare conductor, commonly used in service drops.

Two-way switch   Used to control one or more lights or other electrical outlets from one location; also called single-pole switch.

Underwriter's knot   A knot used to tie two insulated conductors at the terminals inside an electrical plug; used to relieve strain on the terminal connection.

Underwriter's label (UL label)   A label applied to manufactured devices that have been tested for safety by Underwriters' Laboratories and approved for placement on the market. These labs are supported by insurance companies, manufacturers, and other parties concerned with electrical safety.

Volt   A unit of measurement for electrical pressure (comparable to pounds of pressure in a water system).

Voltage   The electromotive force or potential difference between two points of a circuit, measured in volts, that causes electric current to flow. One volt creates a current of 1 ampere through a resistance of 1 ohm.

Voltage drop   A loss of electric current caused by overloading wires or by using excessive lengths of undersized wire. Often indicated by dimming of lights and slowing down of motors.

Volt-ampere   In an AC circuit, a unit of measurement of electric power equal to the product of volts times amperes. In DC power, 1 volt-ampere equals 1 watt; in AC it is a unit of apparent power.

Watt   A unit of measurement of electric power. (Volts times amperes equals watts of electric energy used.) One watt used for 1 hour is 1 watt-hour; 1,000-watt hours equals 1 kilowatt-hour (the unit by which electricity is metered and sold by utility companies).

Wire   An electrical conductor in the form of a slender rod or cable.

Wire gauge   A standard numerical method of specifying the physical size of a conductor. The American Wire Gauge (AWG) series is most common.

Wire nut   *See* Conductor, solderless.

Wiring diagram   A drawing, in symbolic form, plotting conductors, devices, and connections.

INDEX

U.S./Metric Measure Conversion Chart

	Symbol	**Formulas for Exact Measures** When you know:	Multiply by:	To find:	**Rounded Measures for Quick Reference**		
Mass (Weight)	oz	ounces	28.35	grams	1 oz		= 30 g
	lb	pounds	0.45	kilograms	4 oz		= 115 g
	g	grams	0.035	ounces	8 oz		= 225 g
	kg	kilograms	2.2	pounds	16 oz	= 1 lb	= 450 g
					32 oz	= 2 lb	= 900 g
					36 oz	= 2¼ lb	= 1000 g (1 kg)
Volume	tsp	teaspoons	5.0	milliliters	¼ tsp	= 1/24 oz	= 1 ml
	tbsp	tablespoons	15.0	milliliters	½ tsp	= 1/12 oz	= 2 ml
	fl oz	fluid ounces	29.57	milliliters	1 tsp	= 1/6 oz	= 5 ml
	c	cups	0.24	liters	1 tbsp	= ½ oz	= 15 ml
	pt	pints	0.47	liters	1 c	= 8 oz	= 250 ml
	qt	quarts	0.95	liters	2 c (1 pt)	= 16 oz	= 500 ml
	gal	gallons	3.785	liters	4 c (1 qt)	= 32 oz	= 1 liter
	ml	milliliters	0.034	fluid ounces	4 qt (1 gal)	= 128 oz	= 3¾ liter
Length	in.	inches	2.54	centimeters	⅜ in.		= 1 cm
	ft	feet	30.48	centimeters	1 in.		= 2.5 cm
	yd	yards	0.9144	meters	2 in.		= 5 cm
	mi	miles	1.609	kilometers	2½ in.		= 6.5 cm
	km	kilometers	0.621	miles	12 in. (1 ft)		= 30 cm
	m	meters	1.094	yards	1 yd		= 90 cm
	cm	centimeters	0.39	inches	100 ft		= 30 m
					1 mi		= 1.6 km
Temperature	° F	Fahrenheit	⅝ (after subtracting 32)	Celsius	32° F		= 0° C
					68° F		= 20° C
	° C	Celsius	⅝ (then add 32)	Fahrenheit	212° F		= 100° C
Area	in.^{2}	square inches	6.452	square centimeters	1 in.^{2}		= 6.5 cm^2
	ft^2	square feet	929.0	square centimeters	1 ft^2		= 930 cm^2
	yd^2	square yards	8361.0	square centimeters	1 yd^2		= 8360 cm^2
	a.	acres	0.4047	hectares	1 a.		= 4050 m^2